Physical Science

Written by
Pamela Jennett

Editor: Carla Hamaguchi
Illustrator: Darcy Tom
Designer/Production: Moonhee Pak/Andrea Ables
Cover Designer: Barbara Peterson
Art Director: Tom Cochrane
Project Director: Carolea Williams

Reprinted 2009

Table of Contents

Introduction

Each book in the *Power Practice*™ series contains over 100 ready-to-use activity pages to provide students with skill practice. The fun activities can be used to supplement and enhance what you are teaching in your classroom. Give an activity page to students as independent class work, or send the pages home as homework to reinforce skills taught in class. An answer key is provided for quick reference.

The practical activities, charts, diagrams, and definition pages in *Physical Science* supplement and enrich classroom teaching to enhance students' understanding of vocabulary, functions, and processes fundamental to understanding how physical science works. This book features the following topics:
- forces and motions
- chemistry
- atoms and elements
- scientific process
- simple machines
- energy
- light and sound
- magnetism and electricity

Use these ready-to-go activities to "recharge" skill review and give students the power to succeed!

Name ______________________________ Date ______________

The Scientific Method

Scientists study problems and conduct experiments in a variety of ways. However, all use the scientific method. The **scientific method** is an organized way to find answers to a problem. Match each phrase in the word box to an activity that describes it. Then number the descriptions to show the correct sequence for the scientific method.

interpret data	observe and record	make a hypothesis
identify the problem	arrive at a conclusion	test the hypothesis

A ______________________ A group of students discusses what they believe will be the outcome of an experiment they are about to conduct. Each student records a statement that will either be proven or disproven by the experiment. ☐

B ______________________ Each member of the group carefully watches as the experiment proceeds. One group member writes down the comments of the group as they call out what they see. ☐

C ______________________ After the experiment is complete, the group discusses their observations. They review their notes and create a graph that shows the results of the experiment. The group discusses what these findings might mean. ☐

D ______________________ Now that the group has decided on a hypothesis, they are ready to proceed with the experiment. As they work, the group is cautious to test only one variable at a time and to follow all directions carefully. ☐

E ______________________ The group reviews their notes and the data they have collected. After a short discussion, they decide whether or not the original hypothesis is correct. ☐

F ______________________ A science group begins a discussion related to what they have been studying in class. They take turns posing questions they still have about the topic. Together, they decide on an experiment they would like to conduct. They hope the experiment will answer some of the questions they still have. ☐

Name ______________________________ Date ______________

Scientific Definitions

The scientific method uses specific vocabulary related to each step in the process. Match each term in the word box to its definition.

hypothesis	control	variable
experiment	procedure	theory
data	conclusion	

1. ______________ This is the organized process used to test a hypothesis.

2. ______________ This is an educated guess about the solution to a problem.

3. ______________ This refers to the observations and measurements recorded during an experiment.

4. ______________ This is a factor that changes in an experiment. Proper procedure calls for testing only one of these at a time.

5. ______________ This is a set of statements or ideas that explain a group of facts or phenomena.

6. ______________ This is the judgment based on the results of an experiment.

7. ______________ This is a variable that is kept constant in an experiment.

8. ______________ This refers to the series of steps taken in order to carry out an experiment.

Name ______________________________ Date ______________

Laboratory Safety

When conducting experiments, it is important to handle equipment correctly and follow procedures that keep students and equipment safe. Each of the following scenarios illustrates a safety rule that should be observed in a science laboratory. Match each phrase in the word box to the situation that describes it.

wear proper safety equipment	handle glass and sharp materials carefully
keep workspace clean and organized	secure loose clothing and hair
use scientific smelling when needed	keep papers and other flammables away from flame

1. ______________________________ Tomas and Elizabeth prepare to begin today's experiment. Tomas rolls up his sleeves so they will not be in the way. He remembers how distracting it is to be pushing them up all the time. Elizabeth secures her long hair with a hair band, ensuring that her hair will be out of her face and away from the lab materials.

2. ______________________________ Today Lewis will be working with a Bunsen burner to heat materials in a test tube. Before he begins, Lewis clears away all the papers that covered his workspace. He also removes his sweater from his chair back and moves it to a coat hook on the far wall.

3. ______________________________ Alicia collects sensory data from her experiment. She writes down what she sees and hears. She notes that the substances also produced heat. Alicia is careful to hold the beaker away from her face and fans the air toward her nose as she sniffs carefully. She notices a faint smell like rotten eggs. She is thankful she did not take a big sniff.

4. ______________________________ Robert carries the beaker full of water with both hands, just to be safe. He would not want to clean up the mess or risk being cut if he were to accidentally drop it.

5. ______________________________ The students mop up spills and properly dispose of used materials as they work. In this way they will not contaminate the results of their experiments.

6. ______________________________ Patricia ties on a long apron and dons a pair of plastic safety glasses. The apron keeps materials from staining her clothes. The safety glasses protect her eyes from any flying substances or splashing liquids

Name ______________________ Date ____________

Equipment in the Laboratory

To assist scientists and researchers in their investigations, specialized laboratory equipment serves various purposes. Use the terms in the word box to label the illustrations.

beaker	graduated cylinder	test tube
funnel	tongs	ring stand
test tube clamp	Bunsen burner	balance
Erlenmeyer flask		

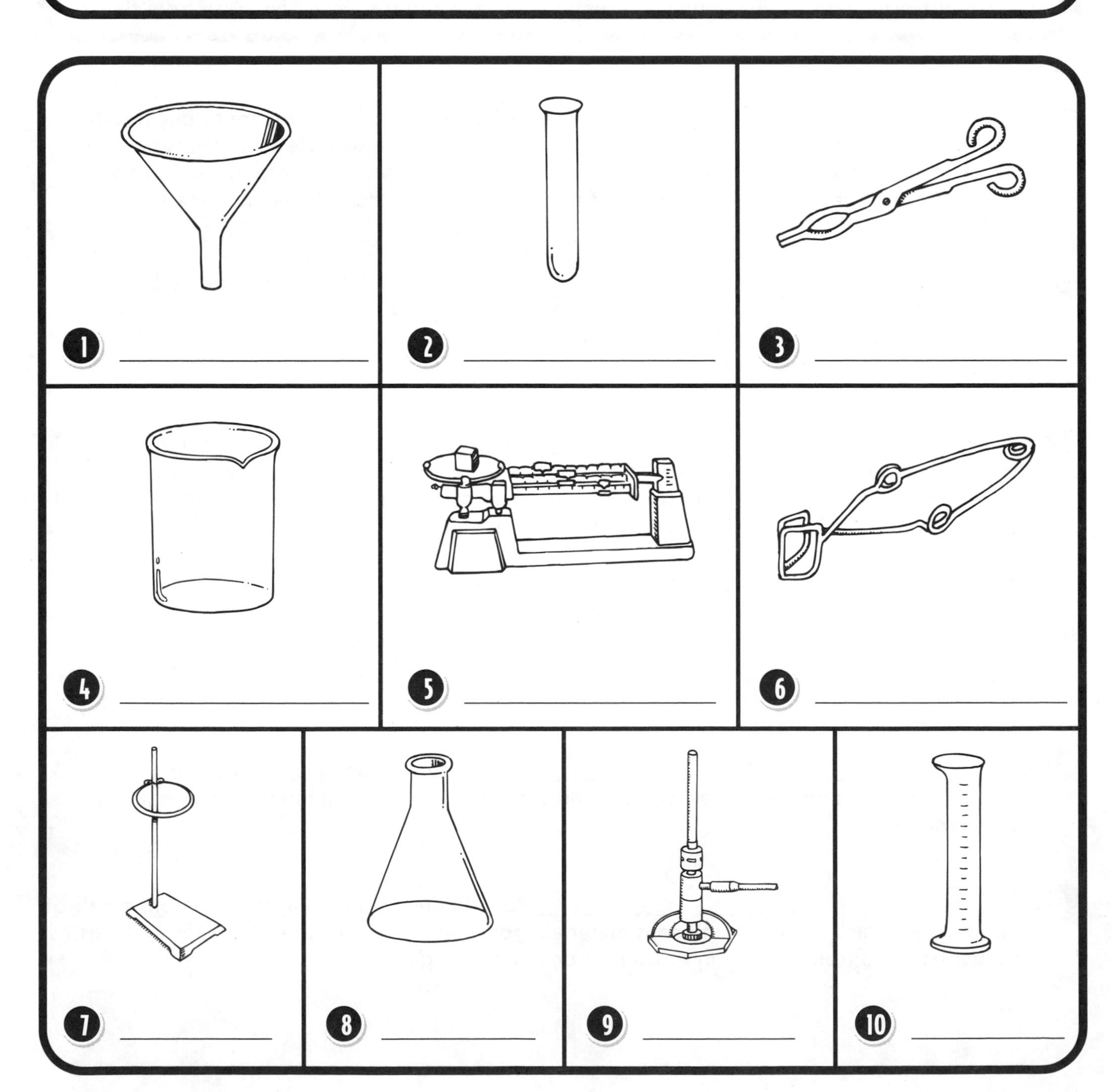

Name ______________________________ Date ______________

Using Lab Equipment

The equipment in a science lab is made to be used safely and for specific purposes. Match each term in the word box to its definition.

beaker	graduated cylinder	test tubes
funnel	tongs	ring stand
test tube clamp	Bunsen burner	balance
Erlenmeyer flask		

1. ______________ A heat-resistant glass vessel used for measuring, storing, or mixing liquids. It has a wide base and the narrow neck can be capped with a rubber stopper or cork.

2. ______________ A wide-mouthed glass vessel used for mixing, measuring and boiling. A spout makes pouring of materials easier.

3. ______________ This tool allows handling and holding of test tubes and other small items.

4. ______________ These small, heat-resistant glass vessels are the basis of all lab work. They allow for the mixing and testing of small amounts of material.

5. ______________ Useful in many lab activities, its narrow profile makes it an accurate measuring device for liquids.

6. ______________ This cone-shaped device allows liquids or powders to be transferred into narrow-necked containers.

7. ______________ This device allows hot beakers or other containers to be picked up safely.

8. ______________ This device accurately measures the mass of objects.

9. ______________ This is a gas burner used in laboratories. It allows for the proper mixture of gas and air.

10. ______________ This device allows containers to be suspended over a work surface or a heat source.

Name ______________________________ Date ______________

Tools for Measurement

Each type of measurement requires a certain type of measuring instrument. Use the terms in the word box to label the illustrations. Use the phrases in the word box to identify each tool and its function.

balance	graduated cylinder	thermometer	spring scale
metric ruler	calipers	measures force	measures temperature
measures mass of an object or material		measures volume of liquids	
measures thickness or small distances		measures length	

1

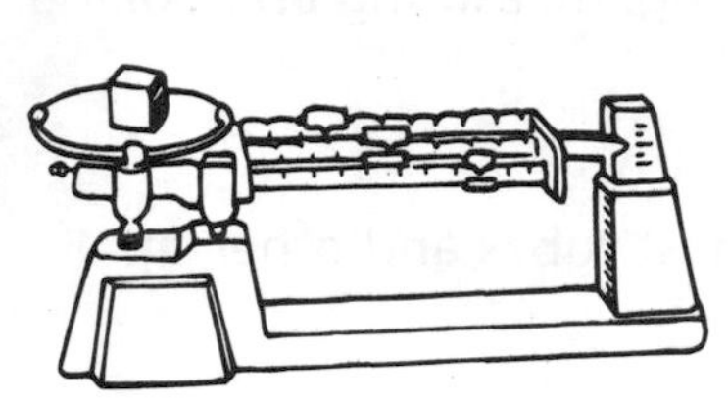

tool: ______________________

function: ______________________

2

tool: ______________________

function: ______________________

3

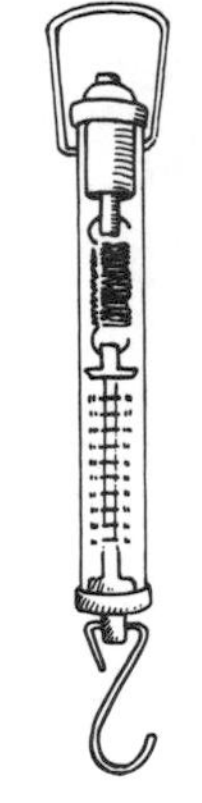

tool: ______________________

function: ______________________

4

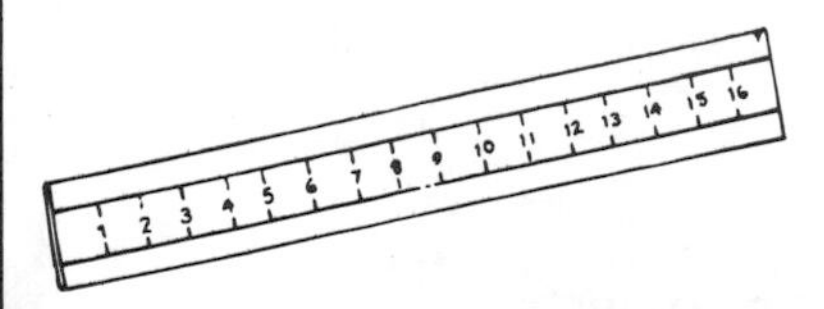

tool: ______________________

function: ______________________

5

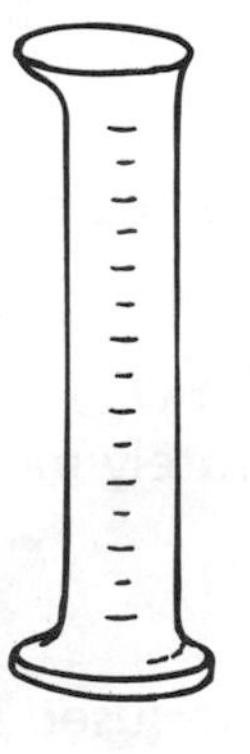

tool: ______________________

function: ______________________

6

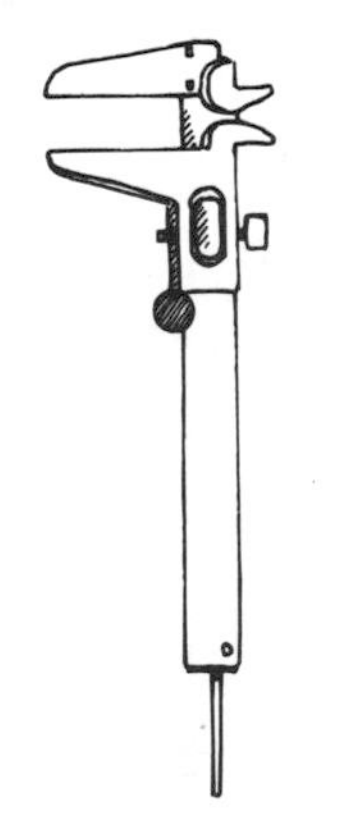

tool: ______________________

function: ______________________

Name ______________________________ Date ______________

Using a Balance Scale

A **balance** is used to measure mass. Use the terms in the word box to label the diagram. Then read the measurement on the diagrams. The balance is measuring in grams.

adjustment screws	riders	pointer
pan	beams	scale

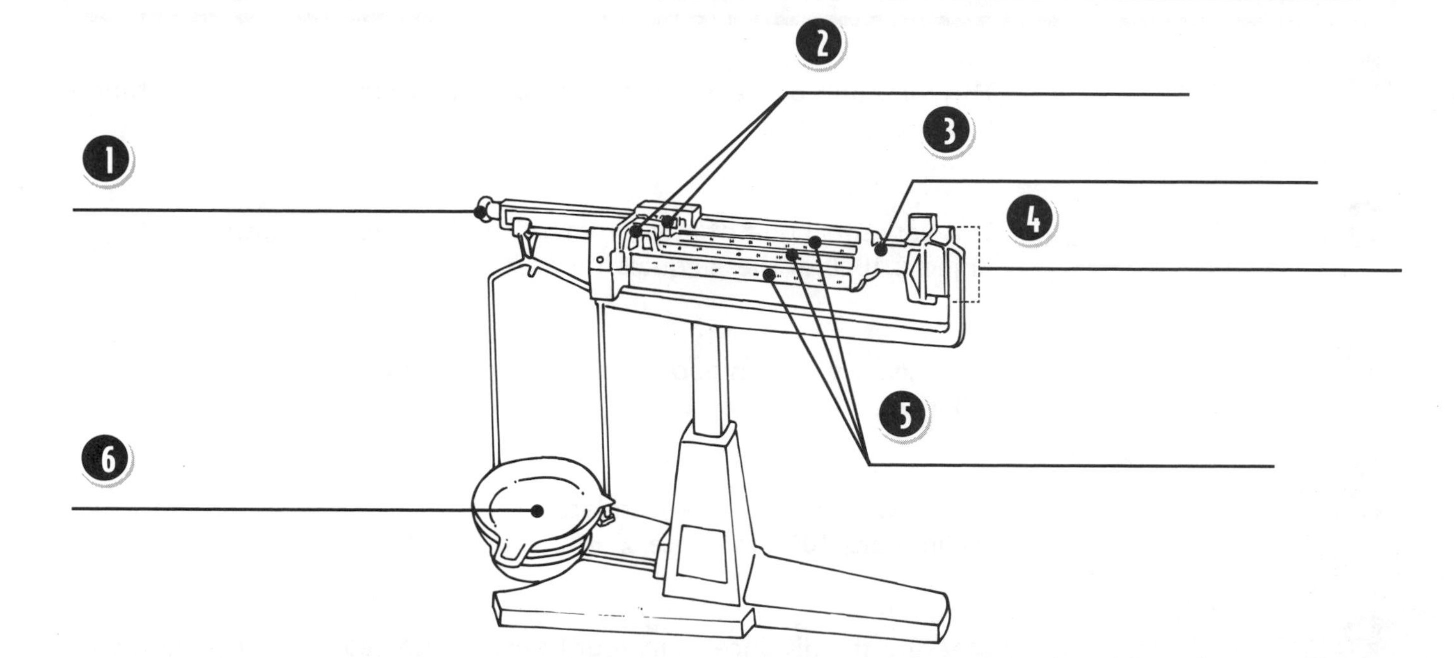

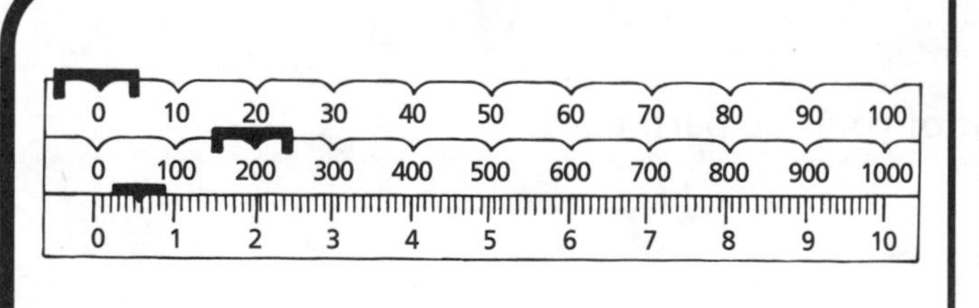

____

____

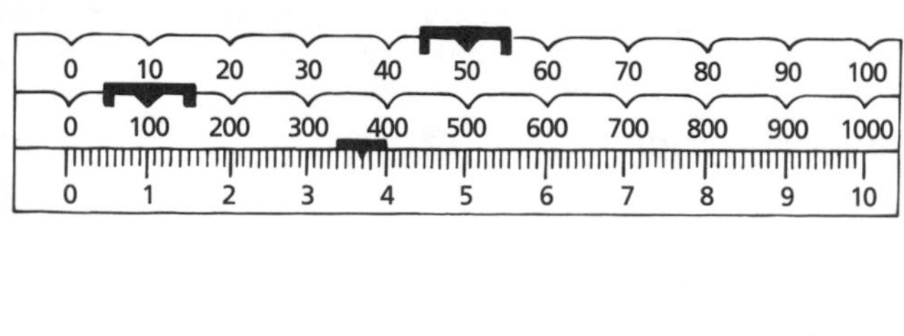

9 ______________________

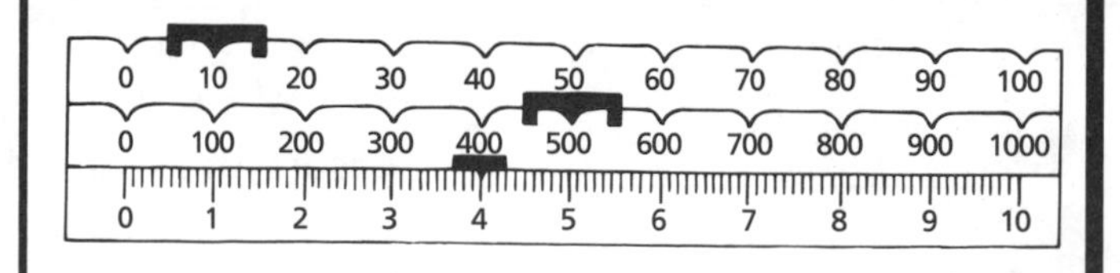

10 ______________________

Name ______________________ Date ____________

Measuring Metric Length

Scientists measure length using the decimal system of measure. The meter is the base of this system. All smaller units are divisions of the meter. All larger units are multiples of the meter. Use the words in the word box to identify the unit descriptions in 1–4. Then write the correct metric measurement to use for each description in 5–10. Some words are used more than once.

millimeter	centimeter	meter	kilometer	decimeter

1. ______________ This long unit of measurement can be used to measure great distances. It is equal to 1000 meters.

2. ______________ This very small unit of measurement can be used to measure tiny objects. 1000 of these equals 1 meter.

3. ______________ This measurement is about the width of your pinkie finger. 10 millimeters is equal to this.

4. ______________ This measurement is the standard for metric linear measurement. 1000 millimeters, 100 centimeters, and 10 decimeters each are equal to this.

5. ______________ Measure the distance from your house to the nearest amusement park.

6. ______________ Measure the height of your lab partner.

7. ______________ Measure the height of a 32-story building.

8. ______________ Measure the width of the common house fly.

9. ______________ Measure the width of your classroom.

10. ______________ Measure the diameter of a marble.

Name ______________________________ Date ______________

Measuring Liquids

Liquids are measured by their volume. The surface tension of a liquid causes it to cling slightly to the sides of the vessel it is in. This slightly bowed surface line is called a **meniscus**. When reading the measurement of a liquid, use the lowest point of the meniscus.

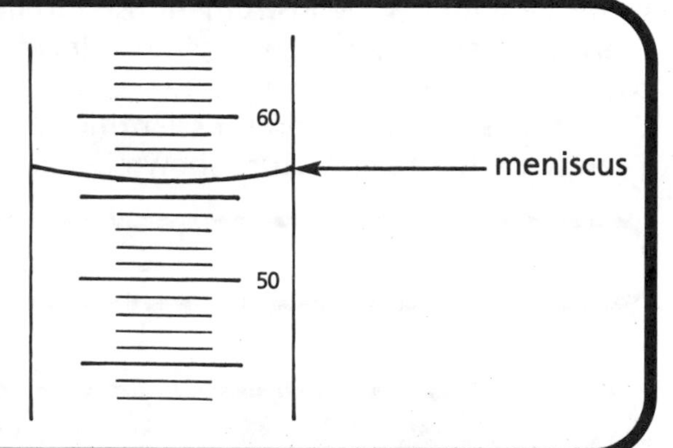

Write the volume shown in each diagram.

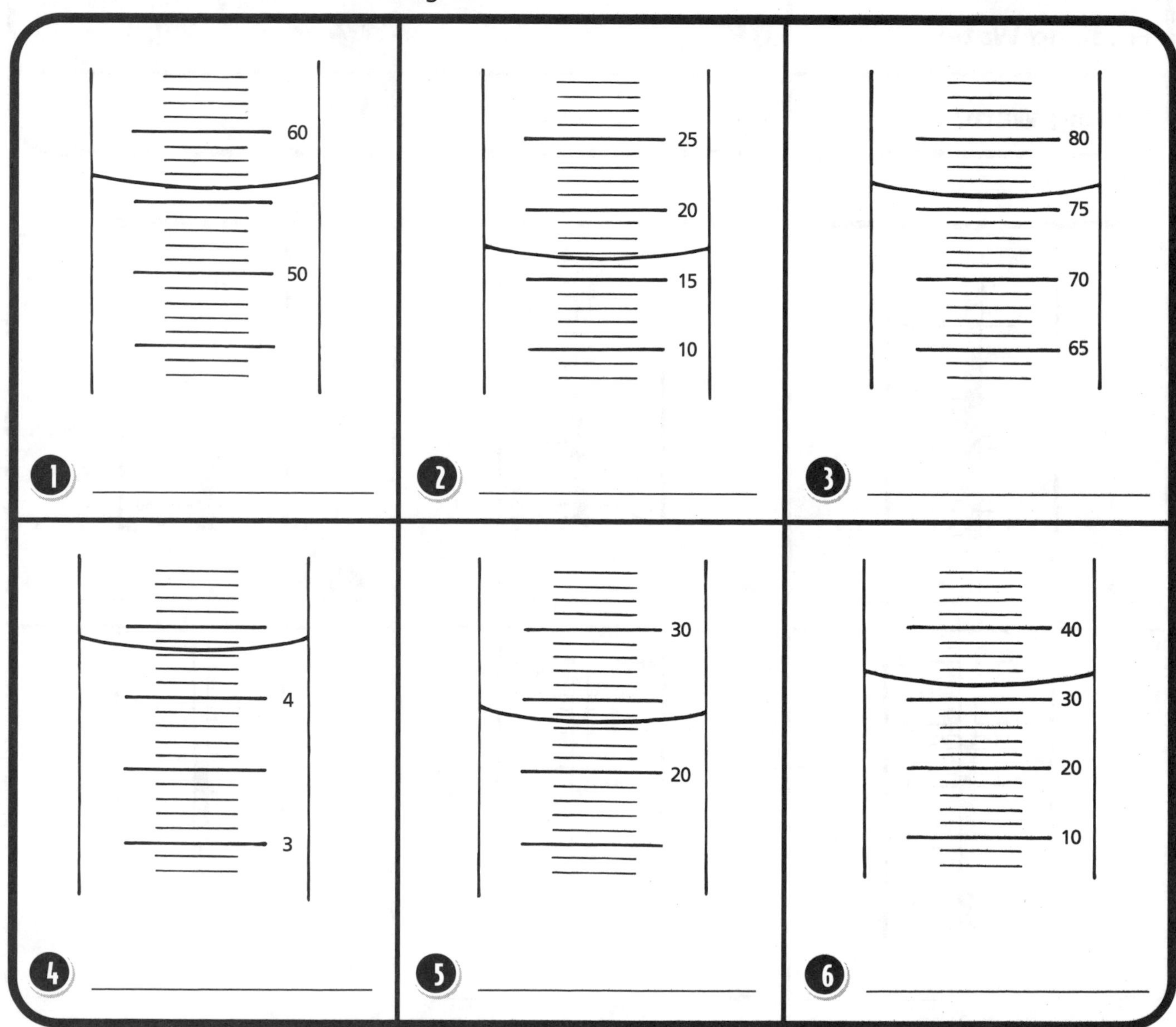

Name ______________________ Date ____________

Measuring Temperature

A **thermometer** is the instrument used to measure temperature. Heat causes the material within the thermometer to expand, showing an increase in temperature. Likewise, cold causes the material to contract, showing a decrease. Two commonly used temperature scales are Fahrenheit and Celsius. Celsius is used most often for scientific purposes. Use the terms in the word box to label the chart. Then write the temperature shown in each diagram.

32 degrees	0 degrees	100 degrees	212 degrees

	Temperature on a Fahrenheit Thermometer	Temperature on a Celsius Thermometer
Freezing Water	1	2
Boiling Water	3	4

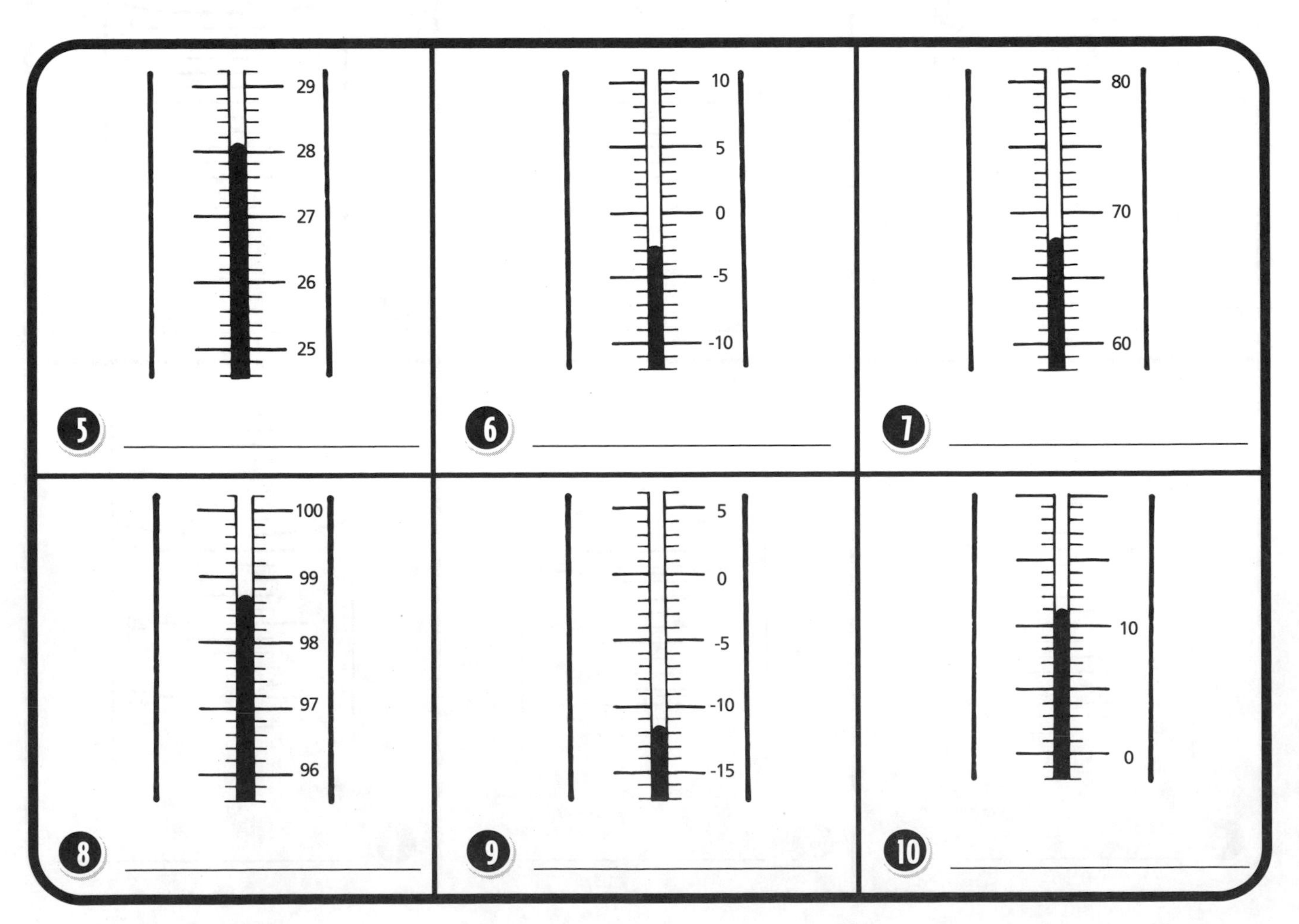

Name ______________________ Date __________

Choosing Units of Measurement

Different substances and objects require different units of measure. For linear measurement, use meter units. For volume, use liter units. For mass, use gram units.

Write the abbreviation for each of the units of measurement.

1 ______nanometer	2 ______millimeter	3 ______centimeter
4 ______meter	5 ______kilometer	6 ______microgram
7 ______milligram	8 ______gram	9 ______kilogram
10 ______millisecond	11 ______milliliter	12 ______liter

Use the abbreviation for each type of unit to label each item with the most appropriate measurement.

Mass of a bowling ball 13 ____________	Mass of a proton 14 ____________	Body mass of an insect 15 ____________
Mass of one paper clip 16 ____________	Distance on a highway 17 ____________	Volume of a bowl of popcorn 18 ____________
Length of a virus 19 ____________	Length of the human intestine 20 ____________	A dose of cough syrup 21 ____________
Length of a blue jay 22 ____________	Length of a beetle 23 ____________	Amount of time light travels a certain distance 24 ____________

Name ______________________________ Date ______________

Scientific Notation

Scientists often work with very small or very large numbers. The number of zeros some of these numbers have can lead to confusion. Scientific notation is used to simplify the number of zeros a number contains. Scientific notation expresses numbers as powers of ten. Use the information in the text box to convert the numbers to scientific notation.

To convert 18,000,000 to scientific notation:

Move the decimal point so there is only one digit to its left, a total of 7 places.

$18{,}000{,}000 = 1.8 \times 10^{7}$

To convert 0.00015 to scientific notation:

Move the decimal point to the right until there is only one digit other than zero to its left, a total of 4 places.

$0.00015 = 1.5 \times 10^{-4}$

Notice that when a number starts out as a fraction less than one, the exponent is always negative.

1. ______________ 5,000,000
2. ______________ 120,000
3. ______________ 22,000,000
4. ______________ 6,000
5. ______________ 275,000,000
6. ______________ 45,000
7. ______________ 0.008
8. ______________ 0.0045
9. ______________ 0.00075
10. ______________ 0.0355
11. ______________ 0.00156
12. ______________ 0.000042

Name ______________________________ Date ______________

The Atom

Atoms are the tiny building blocks of matter. All the matter on Earth is made up of various combinations of atoms. Atoms are the smallest particles of an element that still exhibit all the characteristics of that element. Use the terms in the word box to label the diagram of an atom. Then match each term to its definition. Most terms are used twice.

electron	electron orbit	neutron
nucleus	proton	atom

1 ______________

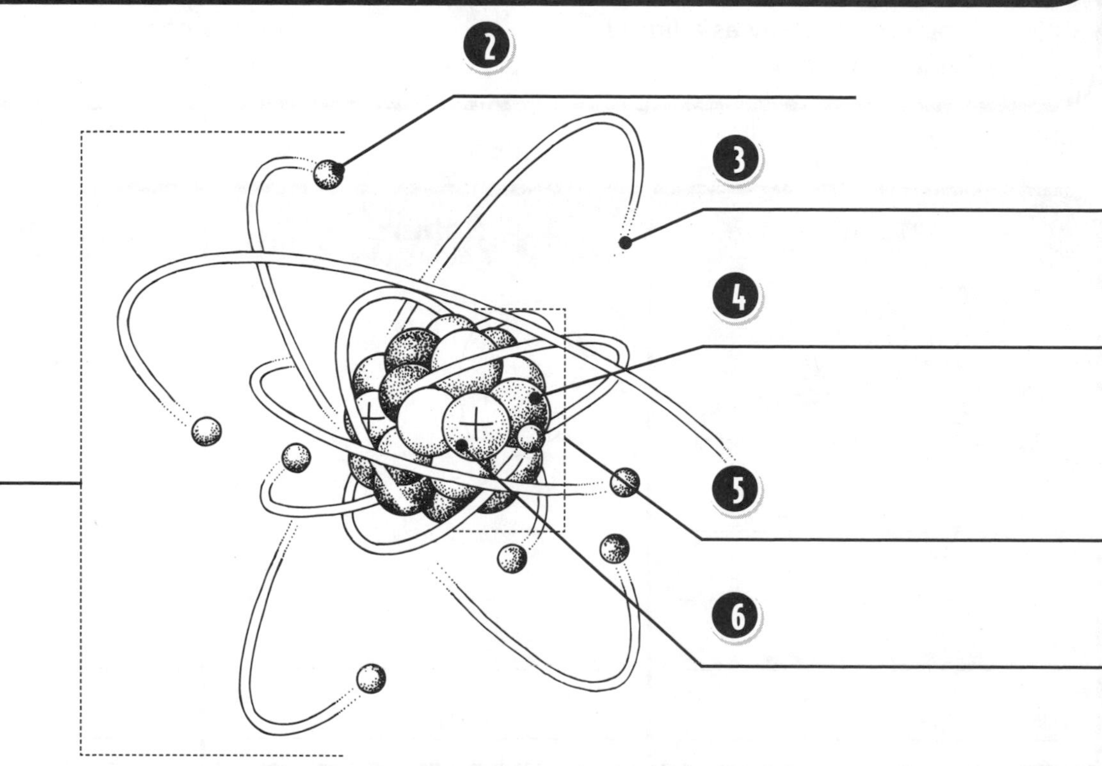

7 ______________ This small particle of an atom carries a negative charge.

8 ______________ Made up of the protons and neutrons, this part of the atom contains nearly all the mass of the atom.

9 ______________ This small particle of an atom carries a neutral charge.

10 ______________ This is the area where electrons travel around the nucleus.

11 ______________ This is the basic building block of all matter.

12 ______________ This small particle of an atom carries a positive charge.

Name ______________________________ Date ______________

Elements

All matter is made up of just over 100 different basic substances called **elements**. Elements cannot be broken down into other substances by heat, light, or electricity. They can be combined to make other kinds of matter. Elements have certain properties that help scientists classify them. Use the phrases in the word box to complete the chart. Some phrases are used more than once.

shiny	conducts heat and electricity
can be hammered into sheets	can be pulled into wires
nonconducting	required for combustion
occurs naturally as a liquid	occurs naturally as a gas
lighter than air	poisonous

Gold 	Helium 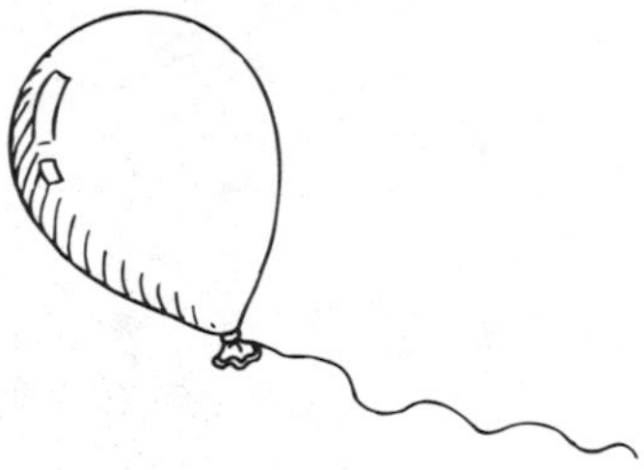	Mercury 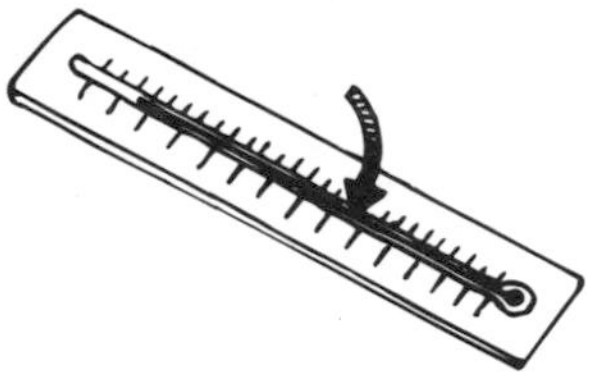
______	______	______
______	______	______
______	______	______
______		

Copper	Silver	Oxygen
______	______	______
______	______	______
______	______	______
______	______	______

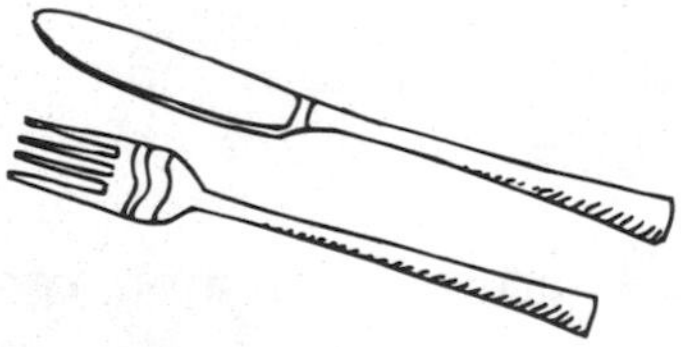

Name ______________________________ Date ____________

Setting the Table

The periodic table is a chart of the chemical elements arranged to show patterns of chemical or physical properties. The elements are arranged on the table based on properties they have in common. Match each term to its definition. You can use the periodic table on page 20 as a reference.

alkali metals	transition	metals
atomic number	alkaline earth metals	naturally
families	noble	periods
rare earth metals		

1. ______________ Elements in the middle of the periodic table are known as these kinds of metals.
2. ______________ These gases are considered inactive. They do not react with other elements.
3. ______________ Most of the elements are considered to be these.
4. ______________ These refer to Group I metals.
5. ______________ Horizontal rows are called this.
6. ______________ The elements are arranged by this.
7. ______________ Vertical columns are called groups or this.
8. ______________ These refer to Group II metals.
9. ______________ For convenience, these are placed at the bottom so the periodic table does not become too wide to be represented in chart form.
10. ______________ There are 92 elements, from hydrogen to uranium that occur in this manner.

The Periodic Table

Period	1A	2A	3B	4B	5B	6B	7B	8B	8B	8B	1B	2B	3A	4A	5A	6A	7A	8A
1	1 H Hydrogen 1.00797																	2 He Helium 4.0026
2	3 Li Lithium 6.941	4 Be Beryllium 9.0122											5 B Boron 10.811	6 C Carbon 12.01115	7 N Nitrogen 14.0067	8 O Oxygen 15.9994	9 F Fluorine 18.9984	10 Ne Neon 20.179
3	11 Na Sodium 22.9898	12 Mg Magnesium 24.305											13 Al Aluminum 26.9815	14 Si Silicon 28.086	15 P Phosphorus 30.9738	16 S Sulfur 32.064	17 Cl Chlorine 35.453	18 Ar Argon 39.948
4	19 K Potassium 39.0983	20 Ca Calcium 40.08	21 Sc Scandium 44.956	22 Ti Titanium 47.88	23 V Vanadium 50.942	24 Cr Chromium 51.996	25 Mn Manganese 54.9380	26 Fe Iron 55.847	27 Co Cobalt 58.9332	28 Ni Nickel 58.69	29 Cu Copper 63.54	30 Zn Zinc 65.37	31 Ga Gallium 69.72	32 Ge Germanium 72.59	33 As Arsenic 74.9216	34 Se Selenium 78.96	35 Br Bromine 79.904	36 Kr Krypton 83.80
5	37 Rb Rubidium 85.4678	38 Sr Strontium 87.62	39 Y Yttrium 88.905	40 Zr Zirconium 91.22	41 Nb Niobium 92.906	42 Mo Molybdenum 95.94	43 Tc Technetium (98)	44 Ru Ruthenium 101.07	45 Rh Rhodium 102.905	46 Pd Palladium 106.4	47 Ag Silver 107.868	48 Cd Cadmium 112.40	49 In Indium 114.82	50 Sn Tin 118.69	51 Sb Antimony 121.75	52 Te Tellurium 127.60	53 I Iodine 126.9044	54 Xe Xenon 131.29
6	55 Cs Cesium 132.905	56 Ba Barium 137.33		72 Hf Hafnium 178.49	73 Ta Tantalum 180.948	74 W Tungsten 183.85	75 Re Rhenium 186.2	76 Os Osmium 190.2	77 Ir Iridium 192.2	78 Pt Platinum 195.09	79 Au Gold 196.967	80 Hg Mercury 200.59	81 Tl Thallium 204.383	82 Pb Lead 207.19	83 Bi Bismuth 208.980	84 Po Polonium (209)	85 At Astatine (210)	86 Rn Radon (222)
7	87 Fr Francium (223)	88 Ra Radium (226.0254)		104 Rf Rutherfordium (261)	105 Db Dubnium (262)	106 Sg Seaborgium (263)	107 Bh Bohrium (262)	108 Hs Hassium (265)	109 Mt Meitnerium (266)	110 Uun Ununnilium (269)	111 Uuu Unununium (272)	112 Uub Ununbium (277)	113 Uut	114 Uuq	115 Uup	116 Uuh	117 Uus	118 Uuo

Note: Elements 113–118 are not currently known. They are shown on the table at their expected positions for information only.

Period															
6	57 La Lanthanum 138.91	58 Ce Cerium 140.12	59 Pr Praseodymium 140.907	60 Nd Neodymium 144.24	61 Pm Promethium (145)	62 Sm Samarium 150.35	63 Eu Europium 151.96	64 Gd Gadolinium 157.25	65 Tb Terbium 158.9254	66 Dy Dysprosium 162.50	67 Ho Holmium 164.930	68 Er Erbium 167.26	69 Tm Thulium 168.934	70 Yb Ytterrbium 173.04	71 Lu Lutetium 174.97
7	89 Ac Actinium (227.0278)	90 Th Thorium 232.038	91 Pa Protactinium (231.0359)	92 U Uranium 238.03	93 Np Neptunium (237.0482)	94 Pu Plutonium (244)	95 Am Americium (243)	96 Cm Curium (247)	97 Bk Berkelium (247)	98 Cf Californium (251)	99 Es Einsteinium (252)	100 Fm Fermium (257)	101 Md Mendelevium (258)	102 No Nobelium (259)	103 Lr Lawrencium (260)

Name ______________________ Date ____________

Decoding the Elements

Each element is represented in a box on the periodic table. Each box gives information about the element. Use the terms in the word box to label the diagram. Then match each term to its definition.

element name	element symbol	atomic number	atomic mass

5 ______________ This represents the element. While the English name sometimes matches, the Latin name was sometimes used to create this.

6 ______________ This is the most commonly used word for the element.

7 ______________ Designated by the number of protons in the nucleus, this determines horizontal placement on the table.

8 ______________ This is determined by the weight of an atom.

Name ____________________ Date ____________

Elements and Their Symbols

Use the periodic table on page 20 to write the common name for each symbol.

1. Cu ____________
2. Ca ____________
3. Pb ____________
4. I ____________
5. K ____________
6. C ____________
7. Sn ____________
8. Ni ____________
9. Al ____________
10. Au ____________

Write the symbol for each element.

11. sulfur ____________
12. flourine ____________
13. xenon ____________
14. tungsten ____________
15. chromium ____________
16. silicon ____________
17. mercury ____________
18. sodium ____________
19. platinum ____________
20. arsenic ____________

Name ______________________________ Date ______________

Real-World Applications

Many of the elements on the periodic table have everyday uses. Match each term in the word box to its description.

tungsten	helium	platinum	nickel	chlorine	chromium
sulfur	neon	lead	tin		

1. ______________ This metal glows when electricity is passed through it. It is used as the filament wire in electric light bulbs.
2. ______________ This shiny metal is used in jewelry and in catalytic converters for cars.
3. ______________ This yellow nonmetal with a distinct odor is used in the manufacture of gunpowder, the vulcanization of rubber, and in insecticides and medicinal drugs.
4. ______________ This is a strong gray metal used in the manufacture of some types of batteries.
5. ______________ This inert gas glows when electricity passes through it, making it useful for displays and signs.
6. ______________ This shiny, mirrorlike metal is mixed with iron to make stainless steel.
7. ______________ In gas form, this element is poisonous. Small amounts in liquid form can purify water by killing bacteria and other organisms.
8. ______________ This gas is lighter than air so it makes balloons rise or float. It is also used as a coolant and in rocket fuel as a pressurizing gas.
9. ______________ This very dense malleable metal is used in plumbing solder, bullets, and shields against radiation.
10. ______________ This silvery metallic metal can be used to coat other metals to protect them against corrosion. It is also used to make alloys of bronze or pewter.

Name ______________________ Date __________

Atoms by Element

An atom's atomic number is the number of protons in its nucleus. Usually an atom has the same number of protons as electrons. The number of neutrons can be different. The electrons form shells around the nucleus of the atom. Each shell holds a specific number of electrons. Use the diagram to guide you as you place the correct number of electrons on each atom illustration.

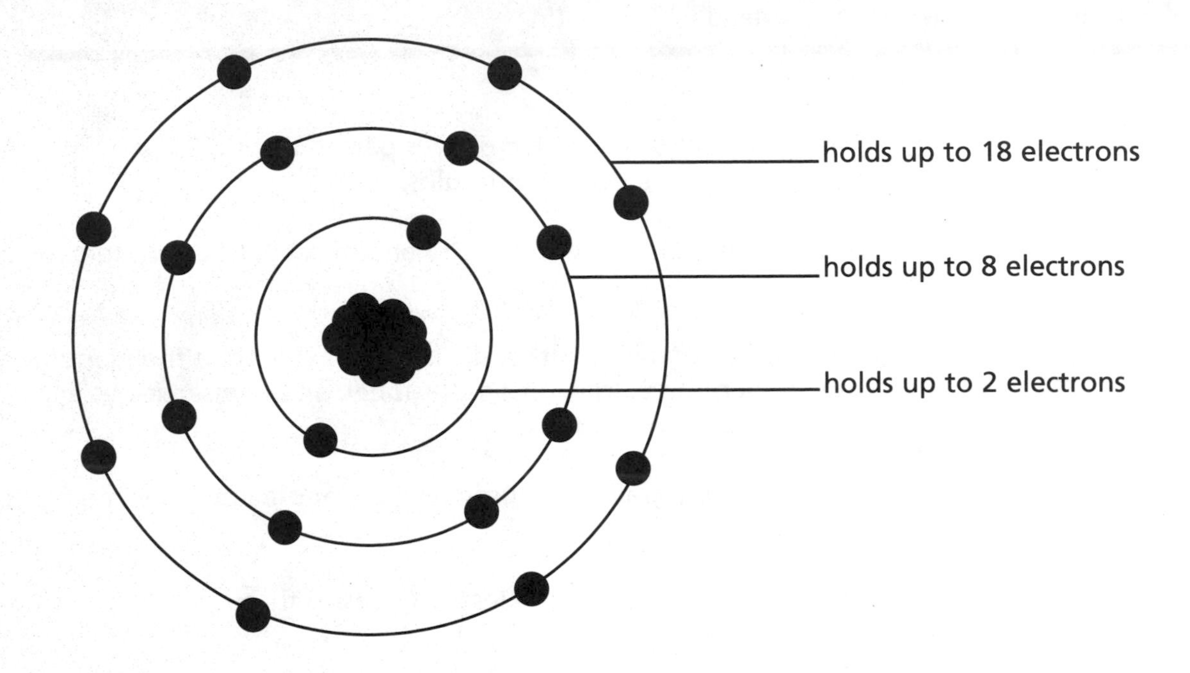

Carbon	Oxygen	Iron
Aluminum	**Nitrogen**	**Calcium**

Name ______________________ Date ____________

Atomic Ions

An **ion** is the name given to an atom or group of atoms that have gained or lost electrons. An atom that loses an electron is positively charged and is called a **cation**. An atom that gains electrons forms a negatively charged ion called an **anion**. Use the periodic table on page 20 to decide if each diagram has lost or gained electrons. Label each as a ***cation*** or ***anion***. Label if each has a ***negative*** or ***positive*** charge.

Sodium

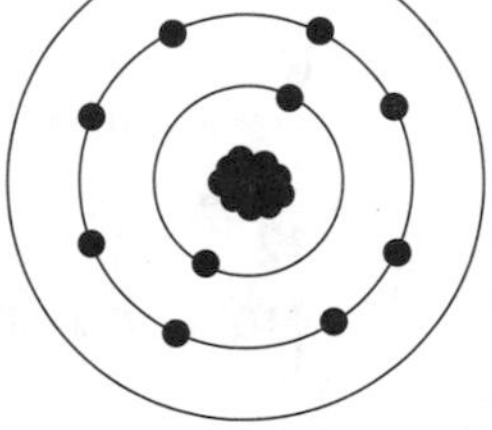

1. Ion: ______________________

   Charge: ______________________

Carbon

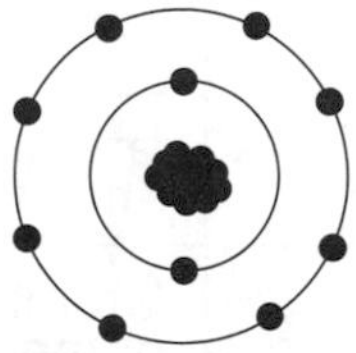

2. Ion: ______________________

   Charge: ______________________

Boron

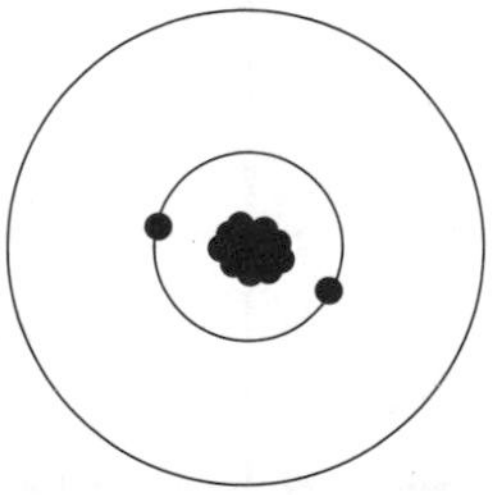

3. Ion: ______________________

   Charge: ______________________

Nitrogen

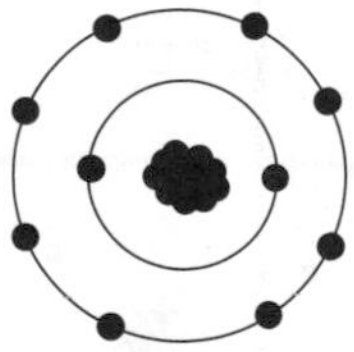

4. Ion: ______________________

   Charge: ______________________

Oxygen

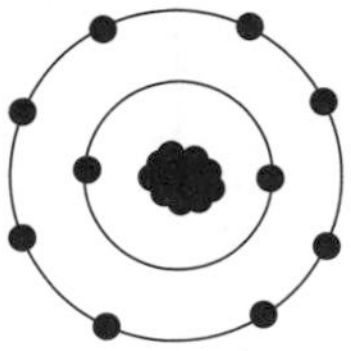

5. Ion: ______________________

   Charge: ______________________

Fluorine

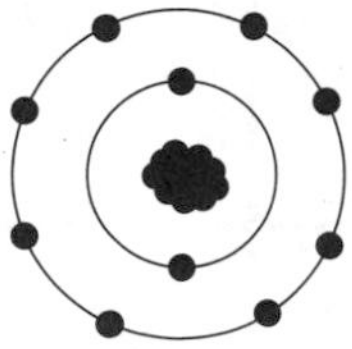

6. Ion: ______________________

   Charge: ______________________

Name ______________________ Date ____________

Symbolic Atoms

A chemical symbol can show information about an atom more easily than a diagram can. The example below shows how to write a symbol for an atom. The atomic number tells how many protons there are. In a neutral atom the electrons equal the number of protons. The mass number equals the sum of the protons and neutrons. If the atom is positively charged, it has lost electrons. If it is negatively charged, it has gained electrons. Use the diagram and the periodic table on page 20 to complete the chart. If there is no charge, assume the atom is neutral.

	Formula	Example with Carbon
# protons	atomic #	6
# neutrons	atomic mass — atomic number	12 − 6 = 6
# electrons	atomic number — number charge	6 − (+4) = 2

atomic number

6

C^{+4} ← change

12

mass number

Use what you have learned from the diagram to complete the chart.

	Element/Ion	Atomic Number	Mass Number	Charge	Protons	Neutrons	Electrons
1	12 Mg 24						
2	9 F^{-1} 19						
3	16 S^{-2} 32						
4	1 H 1						
5	4 Be^{+2} 9						
6	11 Na^{+1} 23						
7	19 K 39						

Name ______________________ Date ______________

Elemental Definitions Part I

Match each element in the word box to its description.

helium	bismuth	phosphorus
carbon	arsenic	selenium
bromine	chlorine	oxygen
fluorine		

1. ______________ This element is found in all organic compounds and in all living organisms.
2. ______________ This nonmetallic element's photoconductive properties make it useful as photocells and solar cells and as a semiconductor.
3. ______________ This highly toxic element is used in medicines and rat poison.
4. ______________ This highly reactive, nonmetallic element occurs naturally in phosphates and is used in safety matches, pyrotechnics, and fertilizers.
5. ______________ This highly corrosive gaseous halogen element is the most reactive of all elements.
6. ______________ This is a nonmetallic element that combines with most elements and is essential for plants and animals.
7. ______________ This is a colorless, odorless, inert gaseous element.
8. ______________ This white, crystalline, brittle metallic element is used in alloys.
9. ______________ This heavy, corrosive, reddish-brown nonmetallic element is used in the production of anti-knock gasolines, dyes, and photographic chemicals.
10. ______________ This highly irritating gaseous halogen can be used as a disinfectant or bleaching agent.

Name ______________________________ Date ______________

Elemental Definitions Part II

Match each element in the word box to its description.

sulfur	nitrogen	neon
cobalt	polonium	zirconium
calcium	lithium	hydrogen
boron		

1. ________________ This is the most abundant element in air.
2. ________________ This is a pale yellow non-metal element found in group 13.
3. ________________ This is a colorless, highly flammable gaseous element, and it is the most abundant element in the universe.
4. ________________ This hard, brittle metallic element resembles nickel and iron in appearance and produces bright blue colors as a salt.
5. ________________ This is a silvery, soft alkaline earth-metal that makes up about 3 percent of the earth's crust and is a basic part of most plants and animals.
6. ________________ A rare, noble gas element, it is colorless but glows reddish-orange when electricity is passed through it.
7. ________________ This is a soft brown metalloid found in group 13. It is important in the field of atomic energy.
8. ________________ A soft, silvery, highly reactive alkali metal, it is used as a heat transfer medium and in batteries.
9. ________________ This is a naturally radioactive metallic element.
10. ________________ This is a silvery-white transition metal used in nuclear reactors.

Name ______________________________ Date ______________

Properties of Metals and Nonmetals

The elements on the periodic table are grouped by metals and nonmetals. Each group has distinct physical and chemical properties. Classify the phrases in the word box to complete the chart.

malleable	lustrous	thallium
gaseous at room temperature	ductile	brittle
forms negative ions	conductor	nonconductor
only forms positive ions	helium	titanium
receives electrons in chemical reactions	covalent bonding	metallic bonding
phosphorus	zinc	boron
selenium	nickel	gold
gives away electrons in chemical reactions	argon	

Properties of Metal Elements	Properties of Nonmetal Elements
______________________	______________________
______________________	______________________
______________________	______________________
______________________	______________________
______________________	______________________
______________________	______________________
Examples	**Examples**
______________________	______________________
______________________	______________________
______________________	______________________
______________________	______________________
______________________	______________________

Name ______________________________ Date ______________

Molecules

A **molecule** is a group of atoms that are tightly bound by chemical bonds. Every molecule has a definite size. If a molecule is broken up into its atoms, these pieces will not behave like the original molecule. A molecule can contain atoms of the same element or atoms of different elements. A substance made up of molecules that include two or more different elements is called a molecular compound. An example of a molecular compound is water. Use the terms in the word box to label the diagrams. Some terms are used more than once.

a water molecule	an ammonia molecule	a carbon dioxide molecule
oxygen atom	hydrogen atom	nitrogen atom
carbon atom		

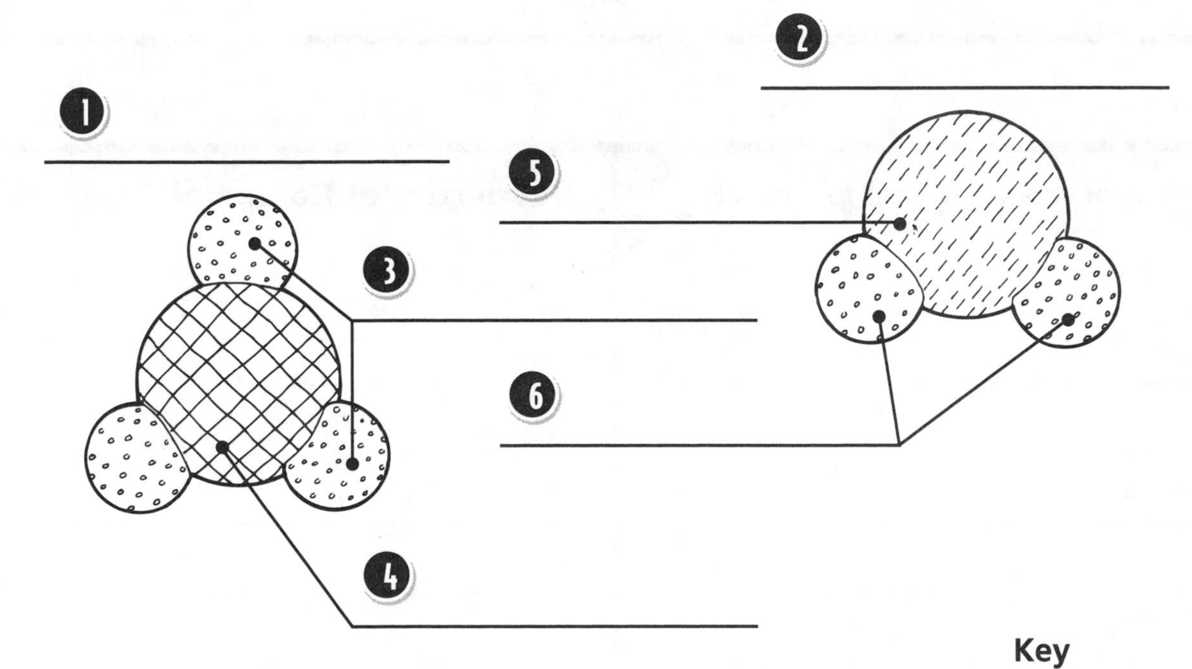

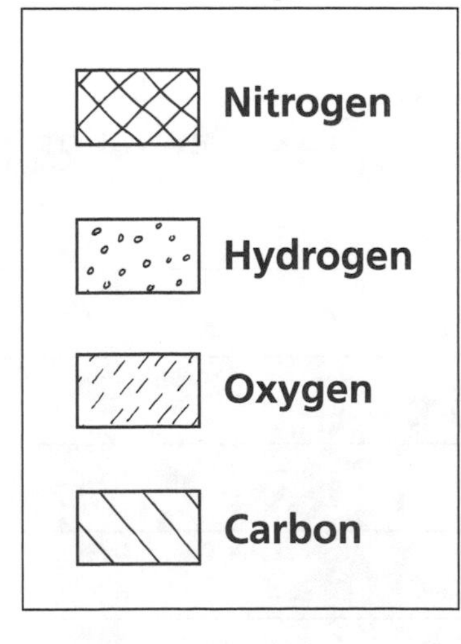

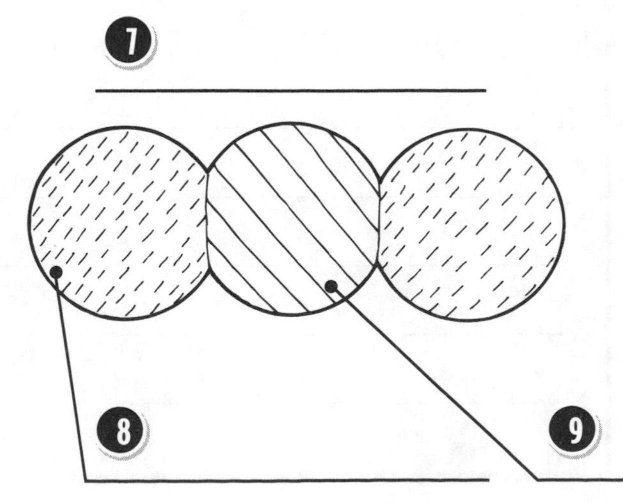

Name ______________________ Date ____________

Three States of Matter

On Earth, matter comes in three common states: solid, liquid, and gas. The state of matter is determined by the strength of the bonds holding its molecules together. Matter can be changed from one state to another through the use of heat. Changes in the three states of matter are physical changes. Classify the phrases in the word box for each state of matter. Some phrases are used more than once.

molecule movement is greatest	has mass	has shape of its own
takes shape of container	has definite volume	has no definite volume
weak bonds between molecules	does not expand	hard to deform
spreads in all directions	expands	takes up space
virtually no bonds between molecules	spreads in direction of gravity	
molecule movement is smallest	strong bonds between molecules	

Solid	Liquid	Gas
____________	____________	____________
____________	____________	____________
____________	____________	____________
____________	____________	____________
____________	____________	____________
____________	____________	____________
____________	____________	____________
____________	____________	____________

Name ______________________ Date ______________

Physical and Chemical Properties

Elements, substances, and compounds have both physical and chemical properties. **Physical properties** are those that can be described using the senses and can be determined without destroying the object. Color, mass, density, and odor are examples of physical properties. **Chemical properties** describe how a substance reacts with another substance and the original is changed into something else. Classify each term in the word box as a physical or chemical property.

reacts with base to form water	density	flammability	solubility
supports combustion	taste	melting point	hardness
boiling point	neutralizes a base	luster	odor
neutralizes an acid	reacts with an acid	reacts to oxygen	electromotive
reacts with water to form gas	color		

Physical Property	Chemical Property
______________________	______________________
______________________	______________________
______________________	______________________
______________________	______________________
______________________	______________________
______________________	______________________
______________________	______________________
______________________	______________________
______________________	______________________

Name ______________________________ Date ______________

Physical Changes

In a physical change, the original substance still exists. It has only changed in form. Match each term in the word box to the description for a physical change.

condensation	sublimation	solid	plasma	density
gas	physical	fusion	evaporation	solution
liquid	substances	suspension	crystallization	

1. ________________ A process by which a gas changes to a liquid.
2. ________________ This describes a unique property in which a substance can go from a solid to a gas without becoming a liquid.
3. ________________ The state that describes where molecules are very close together and are regularly arranged.
4. ________________ A process by which a liquid changes to a gas.
5. ________________ This term describes the mass of a substance divided by its unit volume.
6. ________________ This term describes the physical change of a liquid to a solid right at the melting point.
7. ________________ This is a heterogeneous mixture where particles are large enough to be seen with the naked eye.
8. ________________ A process by which a liquid changes to a solid, forming crystals.
9. ________________ This is a type of mixture in which one substance is dissolved into another.
10. ________________ This is a term used to describe any element or compound.
11. ________________ The state of matter that describes a substance that has definite volume but no definite shape.
12. ________________ This type of property can be observed without destroying the original substance.
13. ________________ The state of matter that describes a substance having no definite volume or shape.
14. ________________ Considered a fourth state of matter, it is a gas with electrically charged particles.

Name ______________________________ Date ______________

Chemical Changes

In a chemical change a new substance is produced. Energy changes always are a part of chemical changes. Chemical changes always involve a physical change as well. Match each term in the word box to the description for a chemical change.

catalysts	reaction	enzyme
reactants	products	base
acid	precipitate	combustion
corrosion	flammable	endothermic
exothermic		

1. ____________________ A reaction in which energy is taken in.
2. ____________________ These are chemicals that speed up chemical reactions.
3. ____________________ These are the substances that are present before the reaction.
4. ____________________ This forms when a soluble substance separates and settles out from a solution due to a chemical reaction.
5. ____________________ This describes a protein that starts or quickens a chemical reaction.
6. ____________________ This is the process in which a metal is destroyed by a chemical reaction.
7. ____________________ This is the burning of a substance in the presence of oxygen.
8. ____________________ This is a reaction that gives out energy.
9. ____________________ This is a process in which one or more substances are changed into others.
10. ____________________ These are the substances that are present after the reaction has taken place.
11. ____________________ This is a corrosive substance with a pH of less than 7.0.
12. ____________________ This describes a chemical that catches on fire easily and burns readily.
13. ____________________ This is a substance that has a pH of more than 7.0.

Name ______________________________ Date ______________

Identifying Physical and Chemical Changes

Read each description and classify it as a physical or chemical change.

1. ____________________ Iron rusts.
2. ____________________ Sodium hydroxide dissolves in water.
3. ____________________ A safety match ignites and burns.
4. ____________________ A cube of ice melts to form a puddle of water.
5. ____________________ Icicles form at the edge of a roof.
6. ____________________ Water is heated and changed into steam.
7. ____________________ Milk goes sour.
8. ____________________ A chocolate bar melts in the sun.
9. ____________________ Acid on limestone produces carbon dioxide gas.
10. ____________________ Vinegar and baking soda react.
11. ____________________ A tea kettle begins to whistle.
12. ____________________ Wood and leaves rot to form humus.

Name ______________________________ Date ______________

Mixtures, Compounds, and Solutions

When substances are mixed together, they can be classified as mixtures, compounds or solutions. Classify the terms in the word box to complete the chart.

milk	humidity	coffee
water	sugar	carbon dioxide
air	ink	alcohol
rubbing alcohol	salt water	brass
blood	sugar water	sand
table salt	ammonia	carbonated soft drink

Mixtures	Compounds	Solutions
A mixture is composed of two or more substances, but each keeps its original properties. They are generally easy to separate.	A compound is a substance formed by the chemical union of two or more elements or ingredients in definite proportion by weight.	A solution is a liquid, called a solvent, that contains a dissolved substance, called a solute.
______________	______________	______________
______________	______________	______________
______________	______________	______________
______________	______________	______________
______________	______________	______________
______________	______________	______________

Name ______________________ Date ____________

Separation of Mixtures

A mixture can be separated by mechanical methods. Match the terms in the word box to each description of a various method of separating mixtures.

weight	magnetism	evaporation	sifting or filtering

A mixture is heated until the water changes from a liquid to a gas. The water vapor can be collected and condensed back into liquid form. The remaining substance has then been separated from the water.  1 ______________________	A mixture is passed through a screen that separates large particles from small particles. The smaller particles pass through while the larger particles are collected on the screen. 2 ______________________
A special machine called a centrifuge spins a mixture to separate substances with different masses. The heavier substance is forced to the bottom and the liquid is siphoned off.  3 ______________________	A magnet is used to separate magnetic materials from those that are not magnetic. 4 ______________________

Use the terms from the word box above to identify the method that could be used to separate each of the following mixtures.

5. ______________ A mixture of oil and water
6. ______________ A mixture of lead and aluminum pellets
7. ______________ A mixture of salt and iron filings
8. ______________ A mixture of sand and gravel
9. ______________ A mixture of sugar and water
10. ______________ A mixture of red blood cells and plasma

Name ______________________ Date ____________

Homogeneous and Heterogeneous Mixtures

Mixtures can be heterogeneous or homogeneous. **Heterogeneous mixtures** are those in which the substances do not spread out evenly. **Homogeneous mixtures** are those in which substances are spread evenly throughout. Homogeneous mixtures can also be solutions or colloids. Use the terms in the word box to complete the chart.

flat soda pop	spaghetti sauce	city air
soil	sugar	aluminum foil
black coffee	sugar water	paint
mayonnaise	alcohol	beach sand
vegetable soup	oil and vinegar salad dressing	
chocolate chip ice cream		

Heterogeneous Mixtures	Homogeneous Mixtures
______________________	______________________
______________________	______________________
______________________	______________________
______________________	______________________
______________________	______________________
______________________	______________________
______________________	______________________
______________________	______________________

Name ______________________________ Date ______________

Solutions, Colloids, and Suspensions

A mixture can be described as a solution, a suspension, or a colloid. Each type of mixture has unique properties. Use the terms and phrases in the word box to complete the chart. Some terms or phrases are used more than once.

large particles	minute particles	medium particles
settles out over time	solute is dissolved into a solvent	
particles remain evenly distributed	murky or opaque	separated by filtration
scatters light	does not settle out	muddy water
transparent	fog	orange juice
salt water	whipped cream	milk
oil and vinegar dressing	sugar water	rubbing alcohol

Suspension	Colloid	Solution
Properties: ____________ ____________ ____________ ____________	Properties: ____________ ____________ ____________ ____________	Properties: ____________ ____________ ____________ ____________
Examples: ____________ ____________ ____________	Examples: ____________ ____________ ____________	Examples: ____________ ____________ ____________

Name ______________________________ Date ______________

Types of Chemical Bonds

A **chemical bond** is an attraction between atoms caused by a sharing or transfer of electrons. There are three main types of chemical bonds: ionic, covalent, and metallic. In addition, chemists often recognize another type of bond called a hydrogen bond. Match each term in the word box to its definition.

ionic bond	properties	cation
metallic bonding	covalent bond	hydrogen bond
single covalent bond	anion	double covalent bond
valence electrons		

1. ____________ These are found in the outermost electron shell of an atom. They are where bonds between atoms are formed.
2. ____________ This results when two atoms share valence electrons between them.
3. ____________ This occurs when one atom gains or loses a valence electron from a different atom.
4. ____________ If an atom gains an electron, it is known as this and has a negative charge.
5. ____________ If an atom gives up an electron to another atom, it is known as this and has a positive charge.
6. ____________ The characteristics that define a molecule or substance.
7. ____________ This type of bonding usually occurs in metals, such as copper.
8. ____________ When two valence electrons are shared by two atoms, it is called this type of bond.
9. ____________ When four valence electrons are shared by two atoms, it is called this type of bond.
10. ____________ A less common type, this is a weak bond between hydrogen and another atom, usually oxygen, fluorine, or nitrogen.

Name ______________________________ Date ______________

Identifying Chemical Bonds

Identify each of the following as compounds with ionic bonds or covalent bonds. Write the elements found in each compound. Use the information in the word box to write the type of bond.

ionic: metal bonded with a nonmetal **covalent:** nonmetal bonded with a nonmetal

1 $NaCl$ Elements: ____________ ____________ Bond: ____________	2 CO_2 Elements: ____________ ____________ Bond: ____________	3 H_2O Elements: ____________ ____________ Bond: ____________
4 K_2O Elements: ____________ ____________ Bond: ____________	5 NaF Elements: ____________ ____________ Bond: ____________	6 CH_4 Elements: ____________ ____________ Bond: ____________
7 SO_3 Elements: ____________ ____________ Bond: ____________	8 MgO Elements: ____________ ____________ Bond: ____________	9 HCl Elements: ____________ ____________ Bond: ____________
10 NO_2 Elements: ____________ ____________ Bond: ____________	11 $FeCl_3$ Elements: ____________ ____________ Bond: ____________	12 P_2O_5 Elements: ____________ ____________ Bond: ____________

Name ______________________ Date ____________

Writing in Code

Chemists use a type of shorthand when they write chemical names. These "codes" use the symbols from the periodic table followed by a subscript number to the right of the symbol that tells how many atoms of the element are present in a molecule. If no number is present, there is only one atom. Use the example to figure out how many atoms are present in each molecule.

Example	Aspirin: a pain-killing molecule
	$C_9H_8O_4$ C = carbon, H = hydrogen, O = oxygen _9_ carbon atoms _8_ hydrogen atoms _4_ oxygen atoms

1	**Nicotine: a poisonous molecule** $C_{10}H_{14}N_2$ ___ carbon atoms ___ hydrogen atoms ___ nitrogen atoms
2	**Trimethylamine: a rotten smell molecule** C_3H_9N ___ carbon atoms ___ hydrogen atoms ___ nitrogen atoms
3	**Aspartame: an artificial sweetener** $C_{14}H_8O_5N_2$ ___ carbon atoms ___ hydrogen atoms ___ oxygen atoms ___ nitrogen atoms
4	**Serotonin: a brain chemical** $C_{10}H_{12}ON_2$ ___ carbon atoms ___ hydrogen atoms ___ oxygen atoms ___ nitrogen atoms

Name ______________________ Date ______________

Number of Atoms by Formula

Use the formulas to determine how many atoms are in each molecule. Use the terms in the word box to label each molecule.

iron oxide	hydrogen peroxide	sodium chloride
mercurous chloride	sulfuric acid	phosphoric acid
potassium carbonate	calcium chloride	ammonium bromide
copper sulfate	sodium sulfite	silver nitrate

Example: CO_2

Atoms: 1 atom of carbon and two atoms of oxygen = 3 atoms

Name: carbon dioxide

1. $NaCl$ Atoms: ______ Name: ______	2. H_2O_2 Atoms: ______ Name: ______	3. Hg_2Cl_2 Atoms: ______ Name: ______
4. Fe_2O_3 Atoms: ______ Name: ______	5. H_3PO_4 Atoms: ______ Name: ______	6. K_2CO_3 Atoms: ______ Name: ______
7. $CaCl_2$ Atoms: ______ Name: ______	8. NH_4Br Atoms: ______ Name: ______	9. $CuSO_4$ Atoms: ______ Name: ______
10. H_2SO_4 Atoms: ______ Name: ______	11. Na_2SO_3 Atoms: ______ Name: ______	12. $AgNO_3$ Atoms: ______ Name: ______

Name ______________________ Date ______________

Chemical Formulas by Name

Prefixes and suffixes are sometimes added to chemical formulas to help indentify them. The prefix indicates how many atoms of each element a formula contains. The suffix "-ide" is added to the second element. Use the prefix key in the word box to help you determine the chemical formula for each compound name.

mono = one	di = two	tri = three
tetra = four	penta = five	

1 carbon monoxide __________	2 sulfur trioxide __________	3 sodium chloride __________
4 carbon tetrachloride __________	5 dinitrogen monoxide __________	6 dinitrogen pentoxide __________
7 carbon dioxide __________	8 phosphorus trichloride __________	9 trimagnesium dinitride __________
10 aluminum iodide __________	11 magnesium bromide __________	12 sulfur trioxide __________
13 phosphorus pentachloride __________	14 aluminum chloride __________	15 dihydrogen monoxide __________

Name ______________________________ Date ______________

Forces and Motion

Force refers to any action or influence that causes an object to move. An object experiences a force when it is pushed or pulled by another object. An object can also experience a force because of the influence of a field, such as gravity or magnetism. Match each term in the word box to its definition.

newton	kinetic	momentum	centripetal	mass	inertia
acceleration	gravity	weight	friction	speed	velocity

1. ______________ A force that resists the movement of one surface against another.
2. ______________ The amount of matter in an object.
3. ______________ The type of force that keeps objects moving in a circle or arc.
4. ______________ The force of attraction that exists between two objects.
5. ______________ The distance covered within a certain unit of time.
6. ______________ The rate at which an object changes its velocity.
7. ______________ The rate at which an object is traveling in a certain direction.
8. ______________ The energy of motion or the energy an object has as a result of its motion.
9. ______________ The tendency of an object to remain at rest if at rest or to continue moving in a straight line if in motion.
10. ______________ A unit of measurement for force.
11. ______________ The amount of force exerted on an object due to gravity.
12. ______________ The product of an object's mass and its velocity.

Name ______________________________ Date ______________

Balanced and Unbalanced Forces

Forces can be balanced or unbalanced. Use the terms in the word box to label the diagrams. One word is used twice.

gravity	table	rest
balanced	unbalanced	accelerate

1. Forces are ____________________.

5. Forces are ____________________.

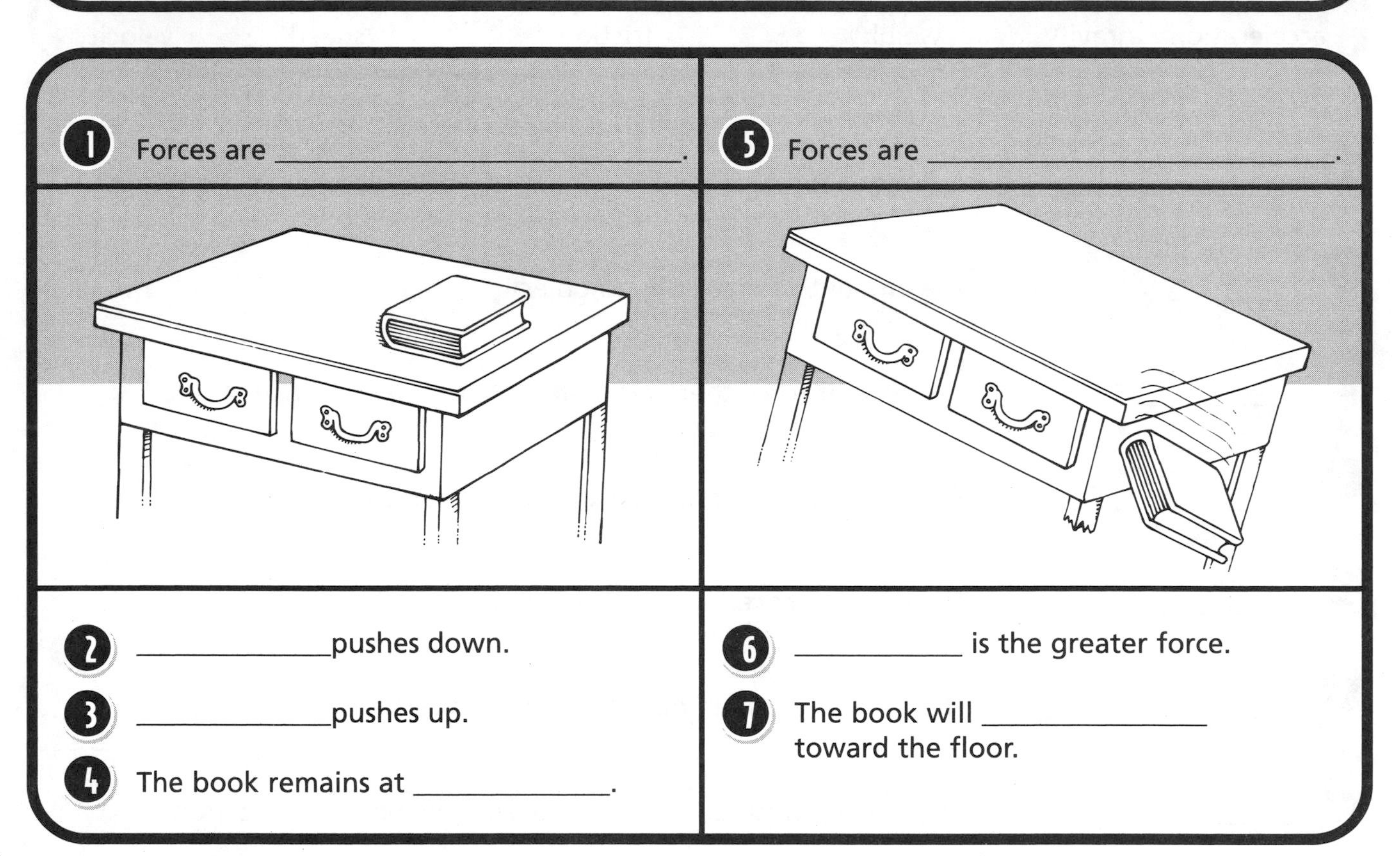

2. ____________ pushes down.
3. ____________ pushes up.
4. The book remains at ____________.

6. ____________ is the greater force.
7. The book will ____________ toward the floor.

Match each term in the word box to its definition.

balanced	equilibrium	inertia	state of motion	zero

8. ____________________ Describes forces working upon an object that are of equal magnitude and come from opposite directions.
9. ____________________ The tendency of an object to resist changes in its state of motion.
10. ____________________ An object at rest has this velocity.
11. ____________________ This describes an object's velocity, or speed with a direction.
12. ____________________ When all the forces acting upon an object balance each other, the object will be in this state.

Name ______________________________ Date ______________

First Law of Motion

Newton's first law of motion states that an object at rest tends to stay at rest and an object in motion tends to stay in motion with the same speed and in the same direction unless acted upon by an unbalanced force. This is also known as **inertia**. Read each description. Then answer the questions.

1. Jonathon wants to put ketchup on his hamburger. He turns the ketchup bottle at an angle toward his plate and smacks the bottom of the bottle until the ketchup comes

What is the unbalanced force? ______________

____

How does inertia affect the ketchup in the bottle? ______________________________

____

2. The Jacksons are driving to the lake when a car in front of theirs slams on its brakes. Mrs. Jackson slams on her brakes, too. Everyone is wearing their seatbelts which stop them from being thrown forward in the car.

What is the unbalanced force? ______________

____

How does inertia affect the each person in the Jackson car? ______________________________

____

3. Terry is riding her skateboard. Suddenly one of the wheels hits a small rock and the skateboard stops. Terry stumbles forward off the skateboard, and catches her balance at the last moment before she falls to the ground.

What is the unbalanced force? ______________

____

How does inertia affect Terry and her skateboard? ______________________________

____

4. Jack is carrying a cup of water. He doesn't see a wrinkle in the floor rug and he trips. Jack manages not to fall or drop the cup, but the water sloshes over the side of the cup and onto the floor.

What is the unbalanced force? ______________

____

What keeps moving? ______________________________

____

Name ______________________ Date ____________

Second Law of Motion

Newton's second law of motion states that acceleration is produced when a force acts on a mass. The greater the mass of the object to be accelerated the greater the amount of force needed to accelerate the object. Each of the following situations demonstrates Newton's second law. Describe how the difference in mass will affect the force needed to change the acceleration..

1. Amy weighs 78 pounds and her dad weighs 187 pounds. They are rollerskating. Amy challenges her dad to a race. They are equally strong. They stand poised at a starting line. Explain who will win?

____

____

____

____

2. Tony and Jose play on the football team. Tony weighs more than Jose. During practice, Tony and Jose practice blocking on a tackle dummy. Both boys start from the same place and position. Each tackle dummy has the same mass. At the same speed, the boys run forward into the dummy. What is their affect on the dummy?

____

____

____

____

3. Two vehicles are broken down on the side of the road. One is a small sports car. The other is a delivery truck. The drivers need to push the vehicles forward and onto the shoulder of the road. Both drivers can push with the same amount of force. Who will get their car off the road first?

____

____

____

____

Name ______________________________ Date ______________

Third Law of Motion

Newton's third law states that for every action, there is an equal and opposite reaction. For each situation, describe the action and the reaction.

1 A boy and girl on roller skates stand facing each other. The girl puts her arms out and pushes away from the boy.	Action: ______________________________ ______________________________ ______________________________ Reaction: ______________________________ ______________________________ ______________________________
2 A golfer swings her club down to hit a golf ball on a tee.	Action: ______________________________ ______________________________ ______________________________ Reaction: ______________________________ ______________________________ ______________________________
3 A frog sits on a lily pad in the middle of a pond. Suddenly it makes a leap pushing off the lily pad.	Action: ______________________________ ______________________________ ______________________________ Reaction: ______________________________ ______________________________ ______________________________

Name ______________________ Date __________

Identifying Newton's Laws

For each illustration, describe how the given laws affect the object's motion.

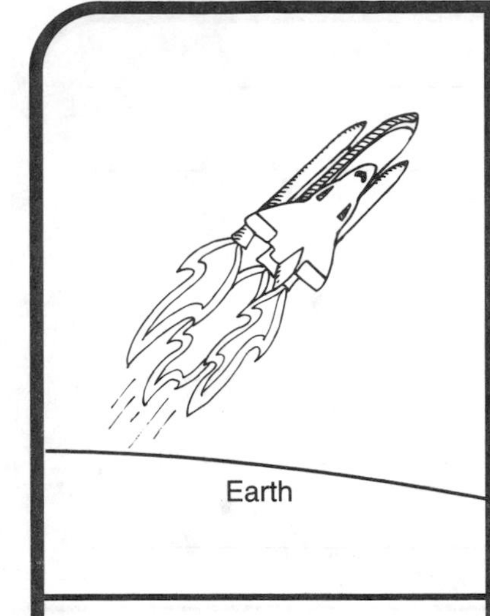

1. First Law of Motion: ______________________

____

2. Third Law of Motion: ______________________

____

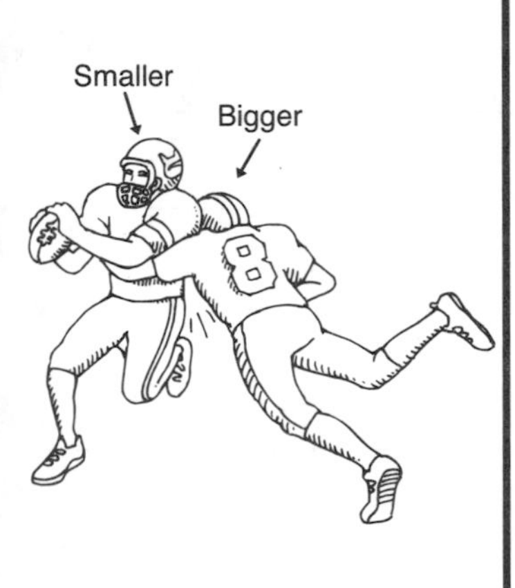

3. Second Law of Motion: ______________________

____

4. Third Law of Motion: ______________________

____

5. First Law of Motion: ______________________

____

6. Third Law of Motion: ______________________

____

Name ______________________________ Date ______________

Determining Speed and Velocity

Speed is a measure of how fast an object is moving. **Velocity** is a measure of how fast an object is traveling in a certain direction. An object can travel at a constant speed that does not change. However, if the direction in which it is traveling does, then its velocity has changed. To find the velocity of an object, use this formula.

$$\text{speed} = \frac{\text{distance}}{\text{time}} \qquad \text{velocity} = \frac{\text{distance}}{\text{time}} \text{ in a specific direction}$$

1. Find the velocity of a truck that travels 75 miles north in 2.5 hours.

   ______________ kilometers per hour

2. Find the speed of a bicyclist who took an hour and a half to travel 10 kilometers.

   ______________ kilometers per hour

3. Find the velocity of a plane that traveled 3,000 miles west in 5 hours.

   ______________ miles per hour

4. Find the velocity of a car that took 7.5 hours to travel 491.25 miles due south.

   ______________ miles per hour

5. Find the average speed of a train that traveled 543 kilometers in 6 hours.

   ______________ kilometers per hour

6. Find the velocity of a train that traveled 420 miles northeast to northwest between two cities in 3.5 hours.

   ______________ miles per hour

7. A plane flies due west for 4 1/2 hours. It travels a total of 5,400 kilometers. What was its velocity?

   ______________ kilometers per hour

8. A cork floats a distance of 8 3/4 miles downriver after a period of 3 hours 30 minutes. What was its average speed?

   ______________ miles per hour

Name ________________________________ Date ______________

Collisions

When two objects collide, the effects depend on the mass and the speed of each object. Ideally, the total momentum of colliding objects is the same before and after a collision. This is referred to as the **Conservation of Momentum.** For each diagram, draw the resulting paths and explain why each outcome occurred.

Head-On Collison	Glancing Blow
Two identical billiard balls	

Why did this happen? ______________	Why did this happen? ______________
A marble and a Ping-Pong ball	

Why did this happen? ______________	Why did this happen? ______________
Two identical marbles	
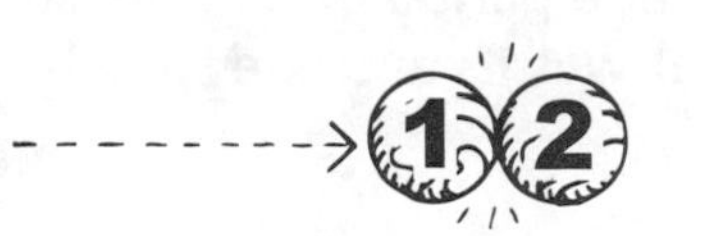	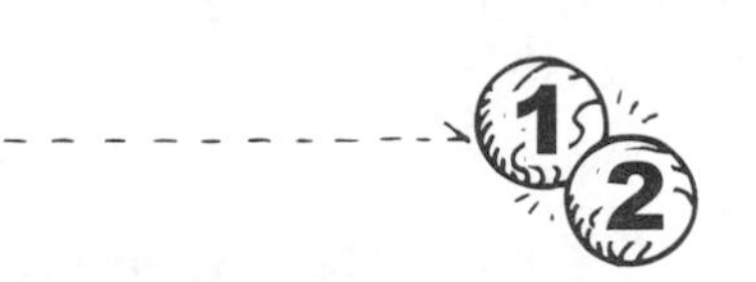
Why did this happen? ______________	Why did this happen? ______________

Name ______________________________ Date ______________

Gravity

Gravity is the force of attraction between two objects. Gravity depends on two factors: the mass of the objects and the distance between them. The Earth is the most massive object near you so you are pulled towards the Earth. The sun is much more massive than the Earth but you don't fly off into space towards it because its distance is so great. Match each gravity related term in the word box to its description.

Newton	acceleration	Einstein	tides	altitude
gravity	inertia	weight	friction	mass

1. ______________ The amount of matter in an object, this characteristic of an object stays the same.
2. ______________ This measure describes the amount of gravitational force of an object.
3. ______________ This is the force of attraction between objects.
4. ______________ The moon is kept in orbit by two forces. Earth's gravity pulls it toward the planet and this force pulls the Moon in a straight line.
5. ______________ The force of gravity was first described mathematically by this English scientist.
6. ______________ A dropped object will increase its speed as it falls toward Earth. If an object held near the surface of Earth is released, it will fall and pick up speed, also known as this.
7. ______________ These phenomenon is caused by the gravitational attraction of the moon and sun on the Earths oceans.
8. ______________ A more accurate theory involving the force of gravity was later formulated by this German-born scientist.
9. ______________ All objects accelerate at the same speed toward Earth. However, this force is the reason that a flat sheet of paper and a crumpled piece of paper do not reach the ground at the same time.
10. ______________ As this increases on Earth, gravitational pull decreases slightly because an object moves farther away from the center of the planet.

Name ______________________ Date ____________

Effects of Gravity

In our daily lives, we use the effects of gravity or overcome the effects of gravity. Study the illustrations and tell how gravity is used to perform the activity. Also tell how gravity must be overcome.

1

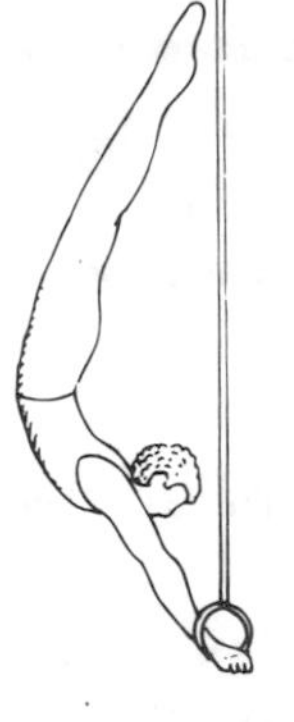

Use gravity: ______________________

____

Overcome gravity: ______________________

____

2

Use gravity: ______________________

____

Overcome gravity: ______________________

____

3

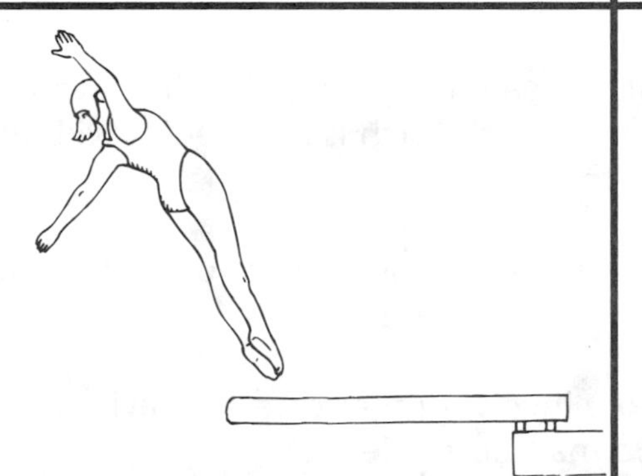

Use gravity: ______________________

____

Overcome gravity: ______________________

____

4

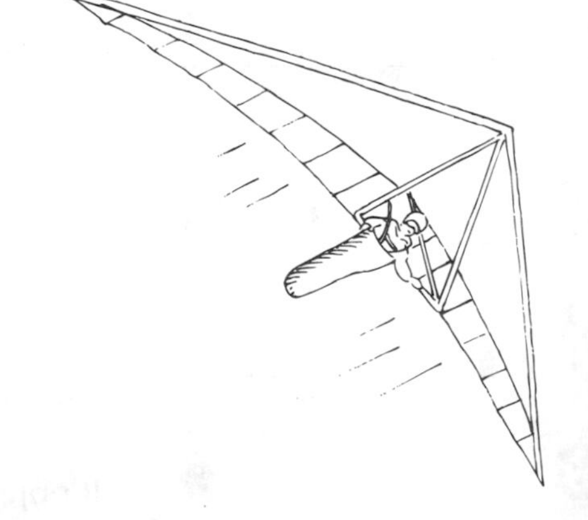

Use gravity: ______________________

____

Overcome gravity: ______________________

____

Name ______________________________ Date ______________

Friction

Friction is a force that acts to slow down moving objects. If friction could be completely removed, then moving objects would continue to move. Friction is a force that opposes the movement of objects sliding over or past each other. Friction can be increased or decreased depending on what needs to be accomplished. Use the terms in the word box to complete the chart.

a canoe glides down a river
grease a bicycle chain
butter a cake pan
oil a door hinge
lotion helps remove a tight gold ring
a dolphin glides through the water
press on a car's brakes
swan dive into a pool
slide down a snowy hill
walk across a wood floor in shoes
rubbing two sticks together starts a fire
grate cheese
pedal your bicycle
build a house of cards
smooth wood with sandpaper
walk across a wood floor in socks
skate across the ice
walk across the road
slide down a grassy hill
rub your hands together to warm them

Friction Is Used	Friction Is Reduced
______________________	______________________
______________________	______________________
______________________	______________________
______________________	______________________
______________________	______________________
______________________	______________________
______________________	______________________
______________________	______________________
______________________	______________________
______________________	______________________

Name ______________________________ Date ______________

Air Resistance

If two objects, one heavy and one light, are dropped simultaneously from a great height, they will both hit the ground at the same time. This is because gravity exerts the same pull on all objects. However, air resistance, or the friction from the air, can affect how things fall. If a wad of paper and a sheet of paper are dropped, the wad will reach the ground first because there is less air resistance. Label each illustration as an example of **Using Resistance** or **Reducing Resistance**. Then explain how you know.

1

____

How? ______________________________

____

2

____

How? ______________________________

____

3

____

How? ______________________________

____

4

____

How? ______________________________

____

Name ______________________________ Date ______________

Potential vs. Kinetic Energy

Potential energy is stored energy. It can be released or harnessed to do work. **Kinetic energy** is the energy possessed by an object as a result of its motion. Label each description as an example of **kinetic energy** or **potential energy**.

1. ______________________ A skier is poised at the top of a steep slope.
2. ______________________ A concrete dam holds back a large reservoir of water.
3. ______________________ An archer has pulled back the string of his bow, ready to release the arrow at the distant target.
4. ______________________ A woman swings her golf club down toward the golf ball sitting on the tee.
5. ______________________ A man swings an axe toward a log.
6. ______________________ A flowerpot is falling from a windowsill.
7. ______________________ A catapult is loaded with a boulder and pulled back into position. It is ready to be launched.
8. ______________________ A fast-moving stream runs toward the mill.
9. ______________________ The baseball player swings her bat.
10. ______________________ A roller coaster has reached the top of the highest crest.
11. ______________________ A marble rolls across the table.
12. ______________________ A child is about to let go of a yo-yo.

Name ______________________________ Date ______________

Calculating Work

Scientifically speaking, work has special meaning. Work is the transfer of energy from one physical system to another. It is expressed by multiplying a force times the distance through which the force moves an object in a certain direction. Work is expressed as joules. Use the formula in the example box to find the work for each situation.

work = force x distance | w = f x d
force is shown in newtons | distance is shown in meters

1. A rock weighing 2 newtons was lifted 3 meters. How much work was done?

   ______________________ joules

2. A rock weighing 6.5 newtons was moved 2 meters. How much work was done?

   ______________________ joules

3. It took 600 newtons of force to move a car 4 meters. How much work was done?

   ______________________ joules

4. It took 45 newtons to lift a crate 1.5 meters. How much work was done?

   ______________________ joules

5. A box weighing 3.2 newtons was moved 2.5 meters. How much work was done?

   ______________________ joules

6. A box weighing 6.4 newtons was moved 2.5 meters. How much work was done?

   ______________________ joules

7. 45 joules were expended to move a box weighing 30 newtons. How many meters was it moved?

   ______________________ meters

8. It took 50 joules to push a crate 2.5 meters. With what force was the crate pushed?

   ______________________ newtons

Name ______________________________ Date ______________

Pendulum Swings

A **pendulum** consists of a suspended object that swings freely on a string. Use the words in the word box to complete each sentence.

frequency	period	bob	gravity
fixed point	equator	simple pendulum	Foucault
Galileo	amplitude	North Pole	metronome

1. A pendulum consists of an object suspended from a ____________________ that swings back and forth.

2. The weighted object at the end of a pendulum is called the ____________________.

3. The angle of each back and forth swing of a pendulum is called a ____________________.

4. The swing of a pendulum is influenced by the force of ____________________.

5. The most basic type of pendulum is called a ____________________ which oscillates back and forth on a single plane.

6. A ____________________ pendulum is used to demonstrate the rotation of Earth.

7. The principle of a pendulum was discovered by the Italian physicist and astronomer ____________________.

8. The back and forth oscillation of a pendulum of a given length remains the same, no matter how large its ____________________.

9. The number of cycles or swings per second is known as the ____________________ .

10. When a pendulum is used to measure Earth's rotation, the effect is most noticeable at the ____________________, where the pendulum rotates every 24 hours.

11. The rotation of a pendulum decreases with latitude. At the ____________________ the pendulum does not rotate at all.

12. A ____________________ uses an upside-down pendulum to keep tempo.

Name ______________________________ Date ______________

The Period of a Pendulum

The period of a pendulum is the time it takes to complete one back-and-forth swing.

If pendula have weights of different masses, the swing periods are still the same.
If pendula have different lengths, the periods are different. The longer the pendula, the longer the swing period.

Graph the information from the table to compare the periods of pendula of various lengths.

Length of pendulum	Number of swings per minute	Length of pendulum	Number of swings per minute
200 centimeters	5	50 centimeters	11
30 centimeters	15	150 centimeters	6
10 centimeters	26	60 centimeters	10
20 centimeters	18	75 centimeters	9
100 centimeters	8	125 centimeters	7

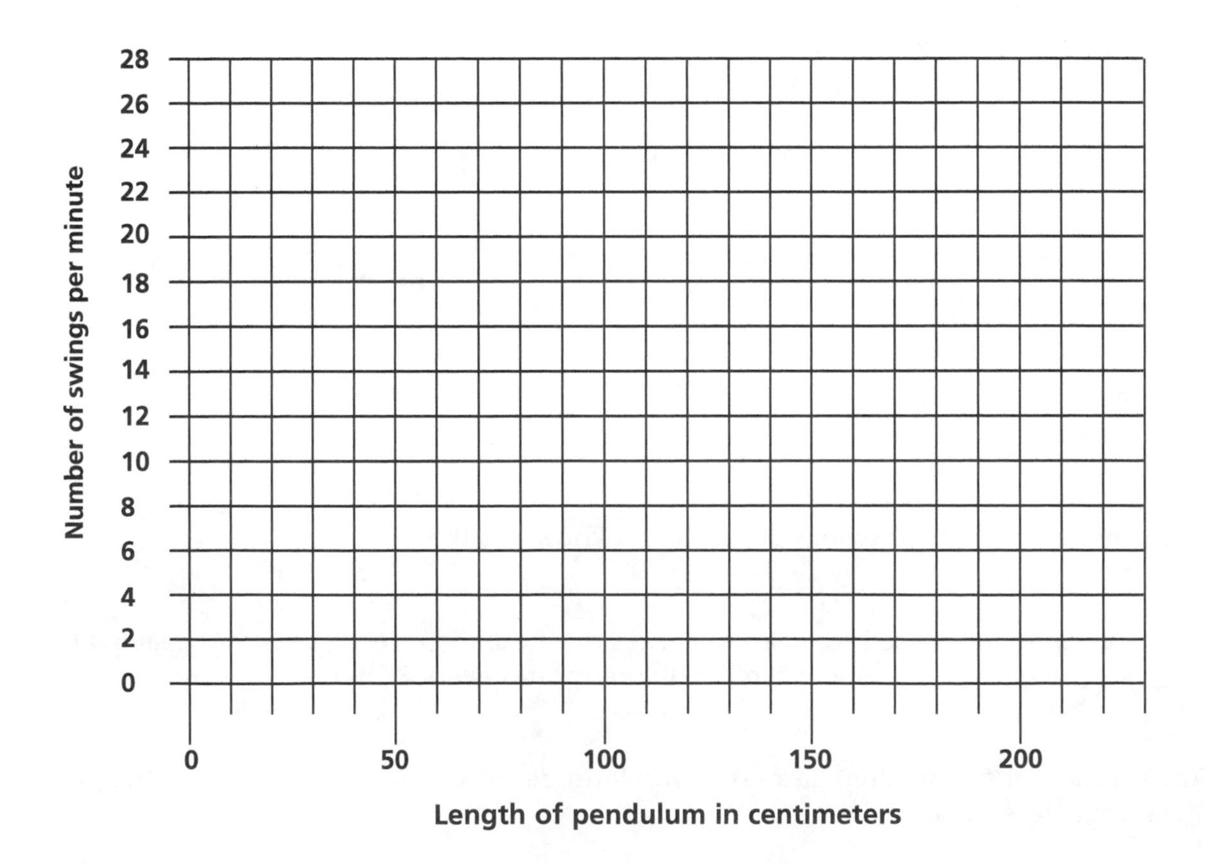

Name ______________________ Date ______________

Simple Machines

A **simple machine** is one that requires only the force of a human to perform work. There are six types. From these types, the elements of all other machines are composed. Use the terms in the word box to label the illustrations. Some terms are used more than once.

lever	wheel and axle	pulley
inclined plane	wedge	screw

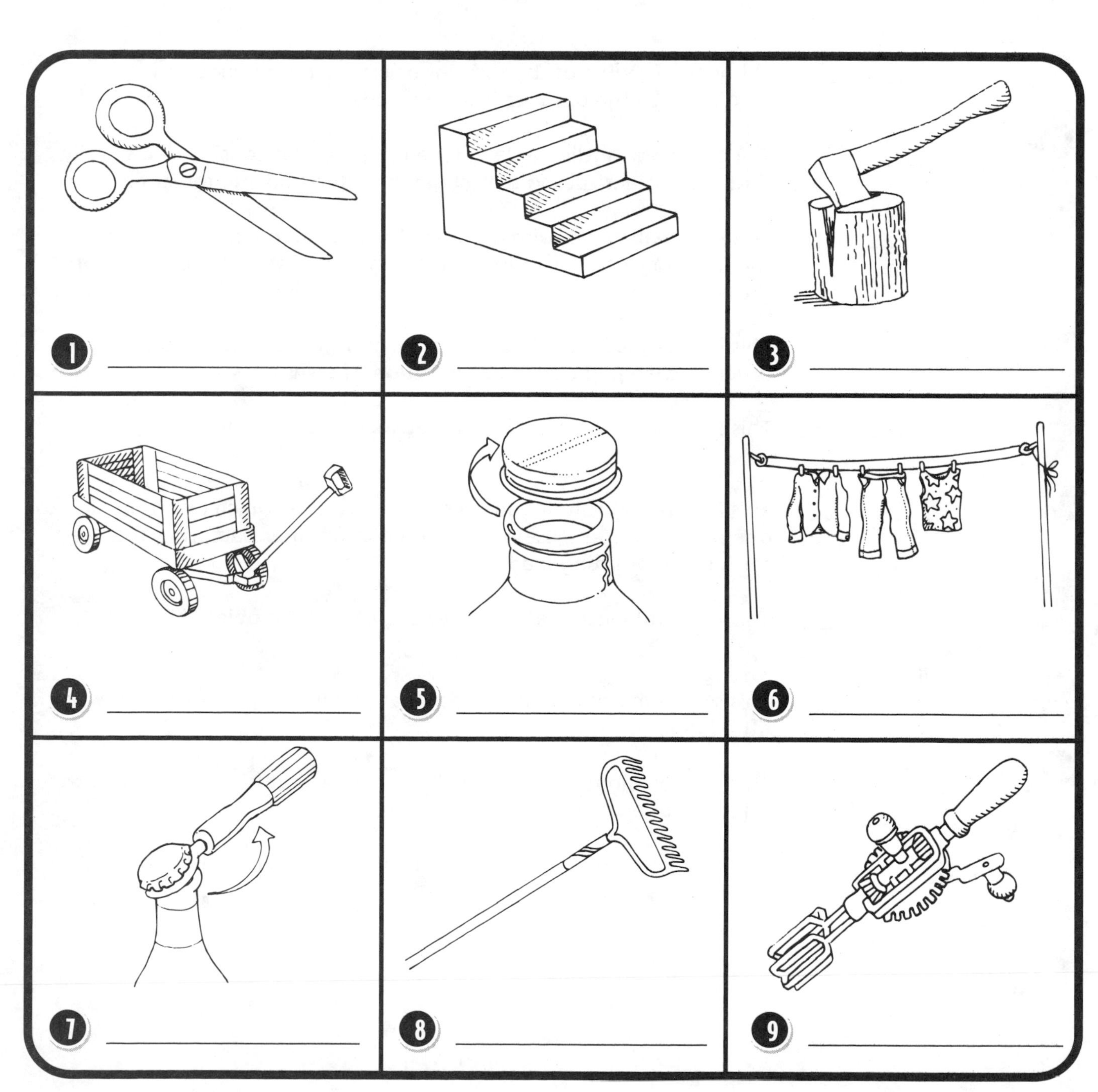

Name ______________________________ Date ______________

Functions of Simple Machines

Simple machines are simple tools used to make work easier. Match each term in the word box to its description.

force	work	distance	lever	inclined plane
wedge	pulley	screw	gear	wheel and axle

1. ______________ This is the product of the force or effort needed to move a load multiplied by the distance it was moved.
2. ______________ This is a simple inclined-plane type machine that consists of a spirally threaded cylindrical rod that engages with a similarly threaded hole.
3. ______________ This is a small wheel with a grooved rim through which a rope or chains run. It changes the direction of a pulling force and combinations of these simple machines increase the force applied for lifting an object.
4. ______________ This is a toothed wheel that engages another toothed mechanism in order to change the speed or direction of transmitted motion.
5. ______________ This is the gap or measurement between two locations.
6. ______________ This is a class of rotating machines or devices in which effort applied to one part produces a useful movement at another part. They are used for moving or lifting loads.
7. ______________ This is an influence that produces a change in an object.
8. ______________ This simple machine is a rigid bar that pivots on a fulcrum to move or lift a load.
9. ______________ This is a combination of two inclined planes that is itself moved with force to cut apart or separate an object.
10. ______________ This simple machine is a slope or ramp that is used to lift a load. It trades distance for force.

Name ______________________ Date ______________

Identifying Parts as Simple Machines

Many of the tools we use every day are based on simple machines. Some tools even have more than one simple machine that makes them work. For example, scissors open and close as a lever but cut through material as a wedge. Use the terms in the word box to label the illustrations. Some illustrations may have more than one term that applies.

lever	wheel and axle	pulley
inclined plane	wedge	screw

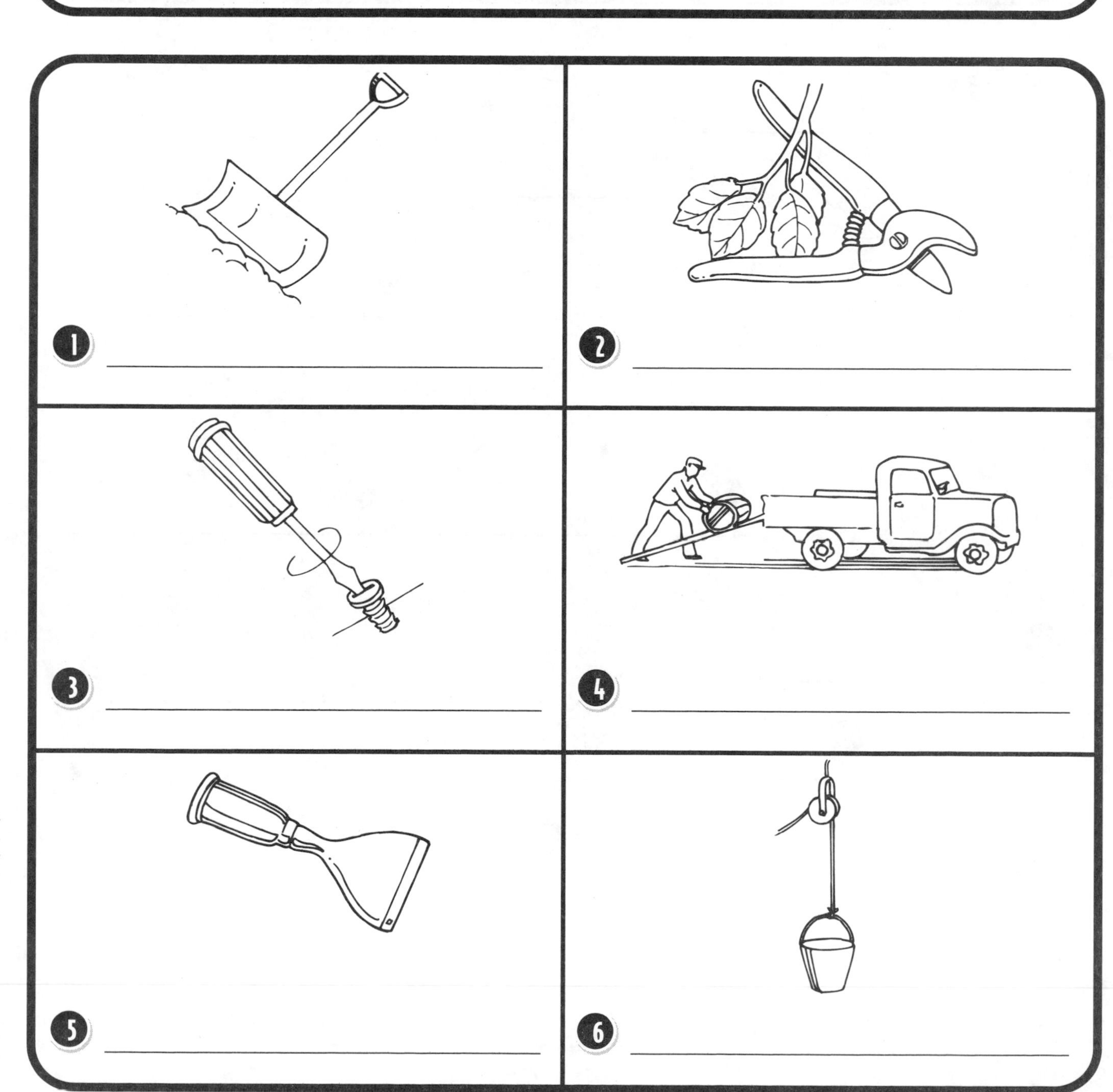

Name ______________________________ Date ______________

Levers at Work

The lever is a simple machine. It is a stiff bar that pivots on a point called a fulcrum. The bar moves but the fulcrum does not. With a lever, a load is lifted a certain distance when you apply force for another distance. Use the terms in the word box to label the illustrations. Some terms are used more than once.

distance you use	distance lever uses	force
load	fulcrum	

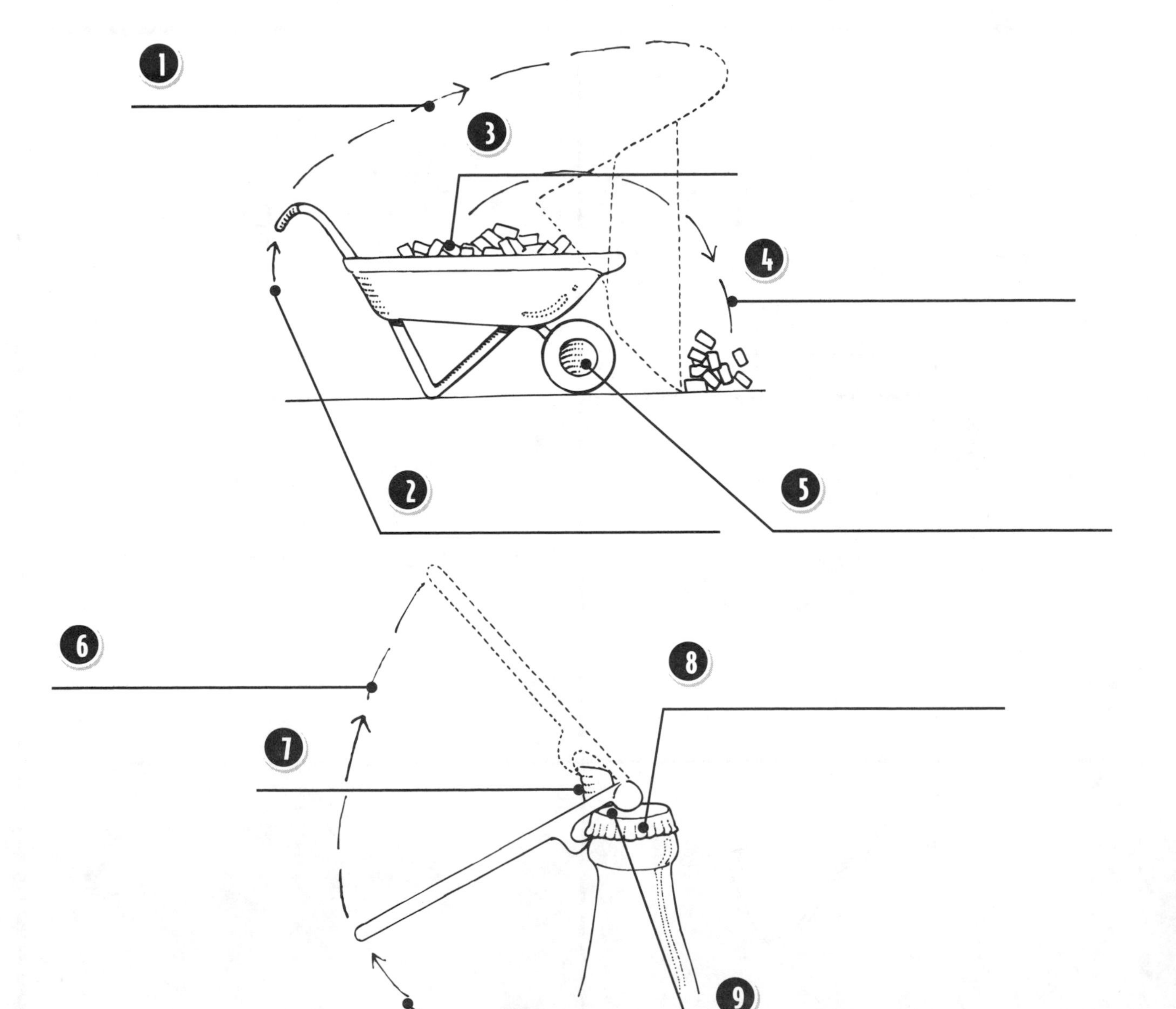

Name ______________________________ Date ______________

Three Classes of Levers

Levers come in three basic classes. They each have a **fulcrum** or pivot point. Each lever has a **force** put into the lever called an **effort** or input force. Each lever also has a force, called the **load**, which is the object being moved. The type of lever is determined by where the effort and load are placed in relation to the fulcrum. Use the terms in the word box to label each class of lever and the diagrams. Some terms are used more than once.

first class	second class	third class
fulcrum	load	effort

Type of Lever: ______________

2
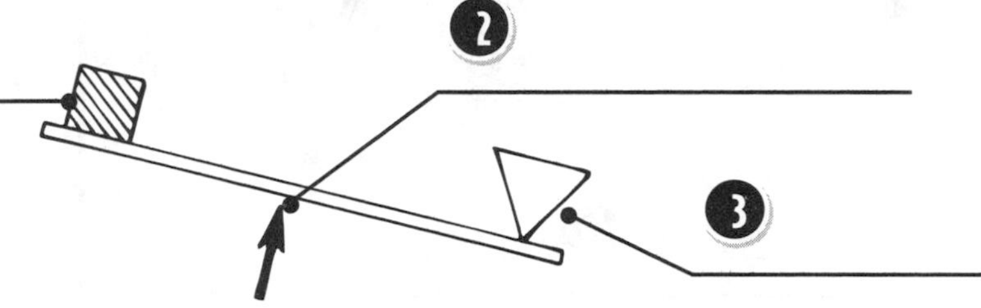
3

The effort and load are on the same side of the fulcrum, but the effort is closer in.

Type of Lever: ______________

5
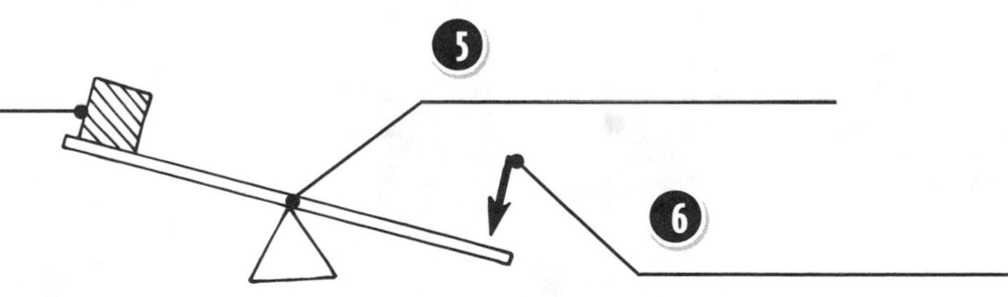
6

The fulcrum is between the effort and the load.

Type of Lever: ______________

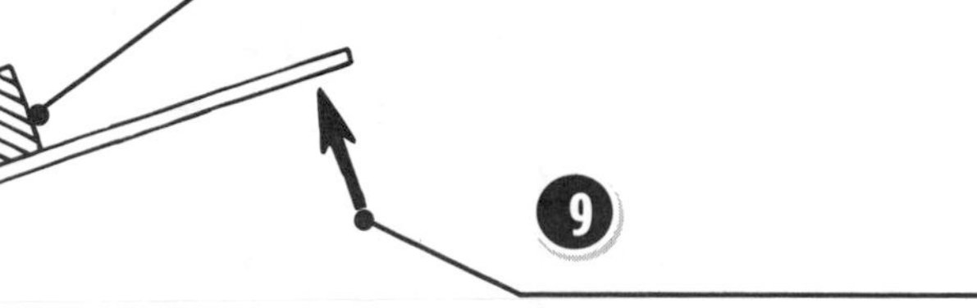
9

The effort and load are on the same side of the fulcrum, but the effort is farther out.

Name ______________________ Date ____________

Classes of Levers

Use the terms in the word box to label each class of lever in the illustrations.

first class	second class	third class

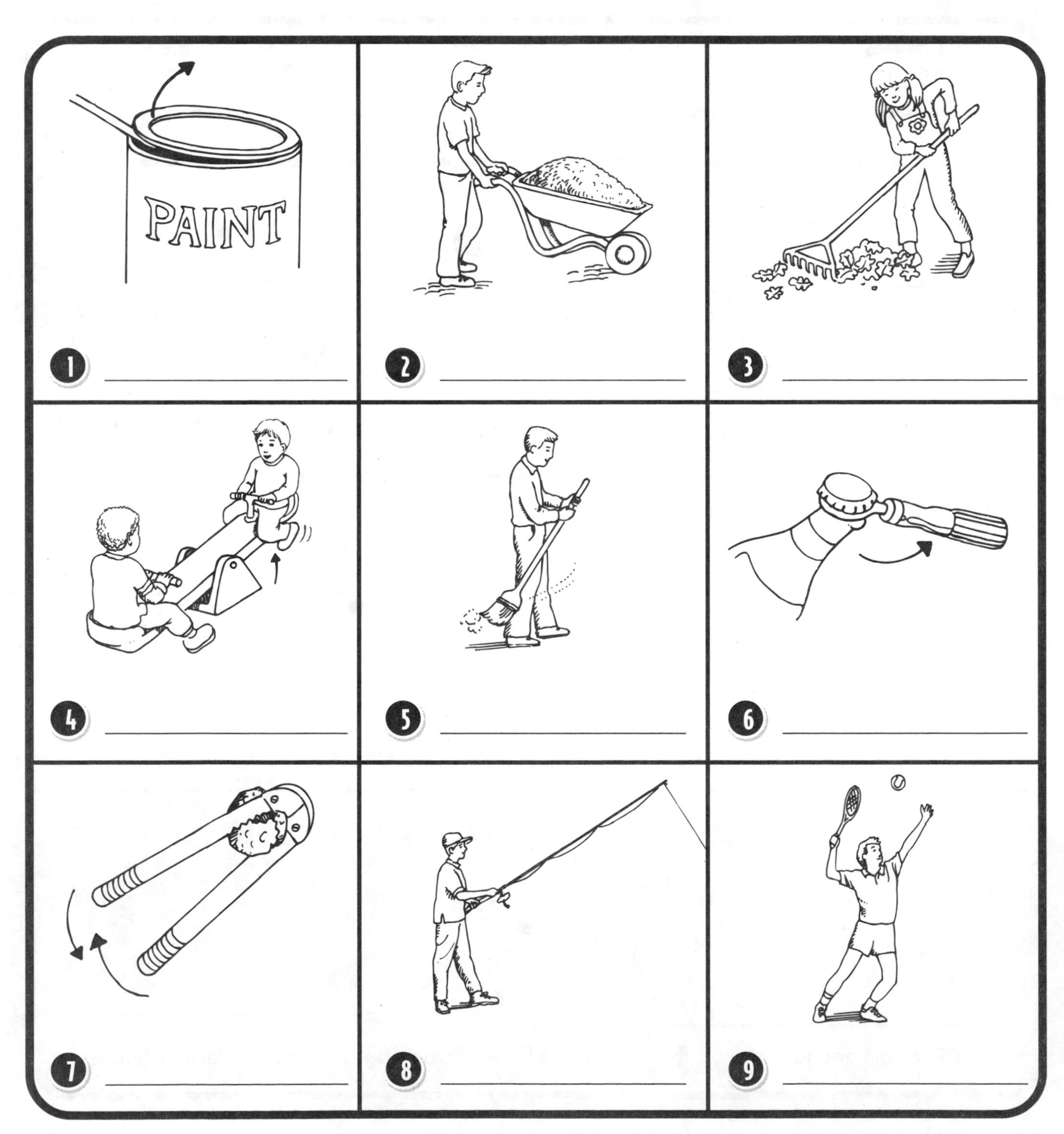

Name ______________________________ Date ______________

Inclined Planes

An **inclined plane** is a slope or ramp that does not move. Instead, it helps you move or raise things that are too heavy otherwise. With an inclined plane, you can do more work with your own force over a greater distance. Describe how the object in each illustration helps you do work more easily.

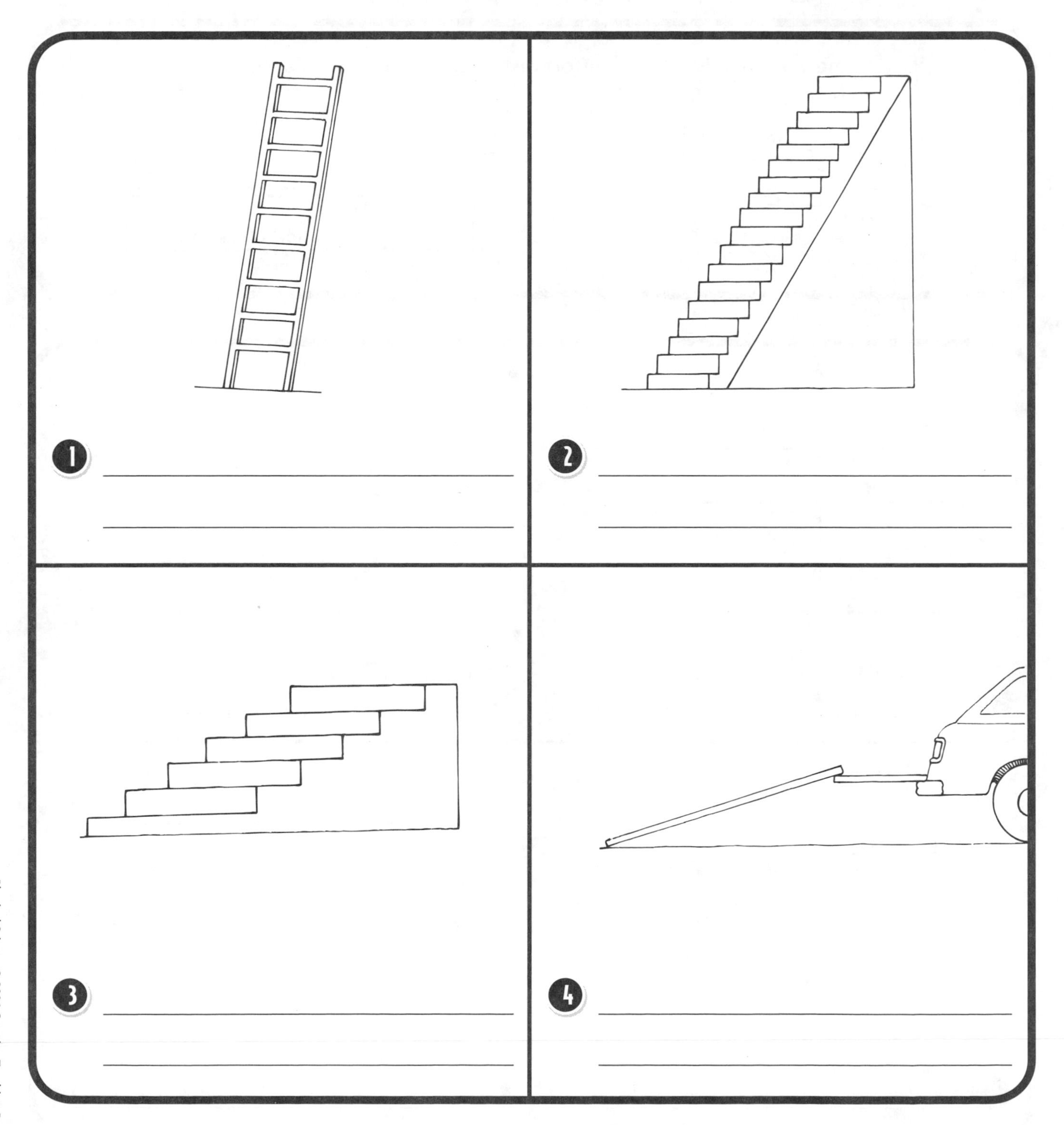

Name ______________________________ Date ______________

Mechanical Advantage of Inclined Planes and Levers

Mechanical advantage is the advantage created by a machine that enables people to do work while using less force. Use the example diagram to identify resistant force and effort force. Then use the formula to calculate the mechanical advantage for each diagram.

mechanical advantage = effort distance ÷ resistance distance

$$MA = \frac{F_E}{F_R}$$

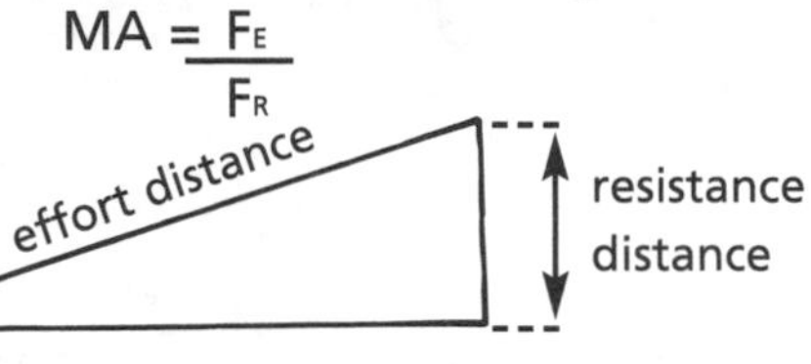

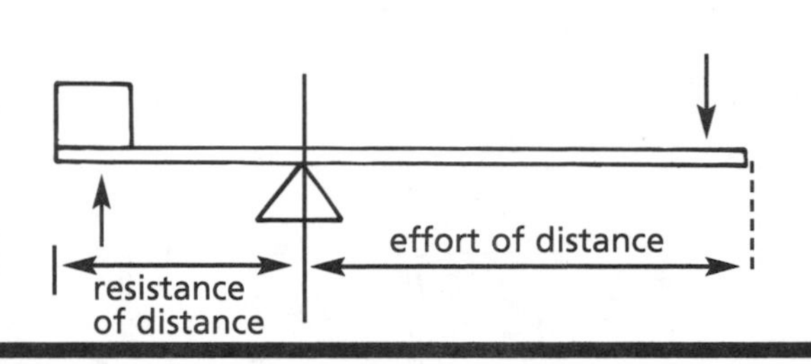

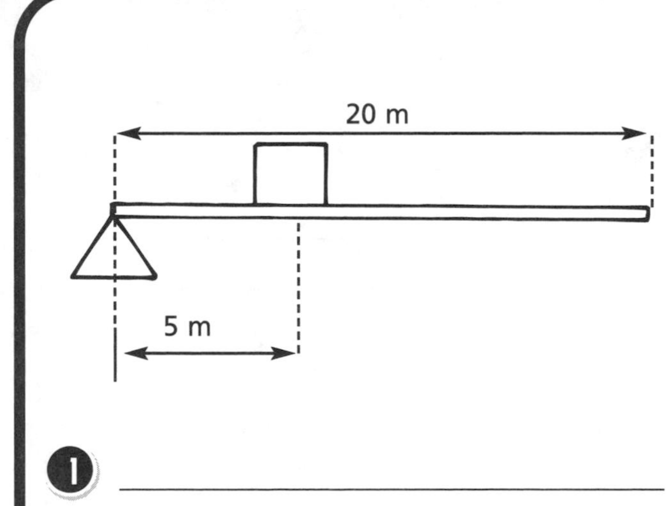

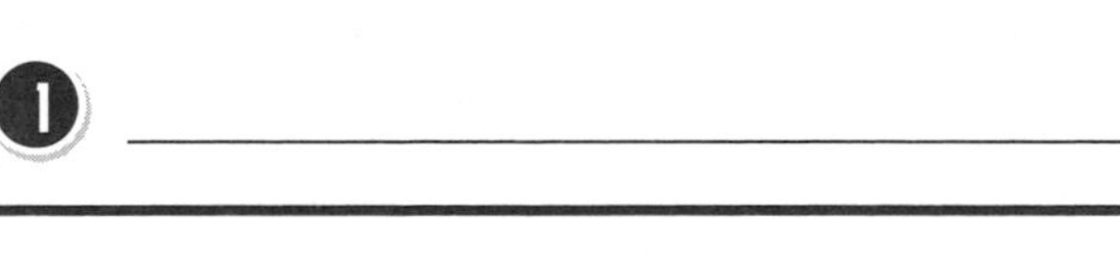

1. ______________________________

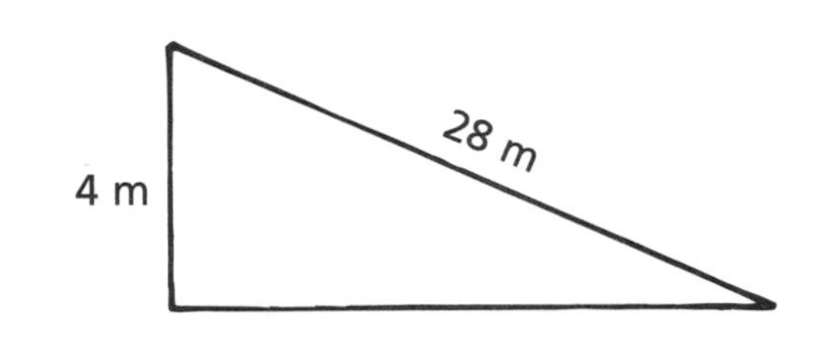

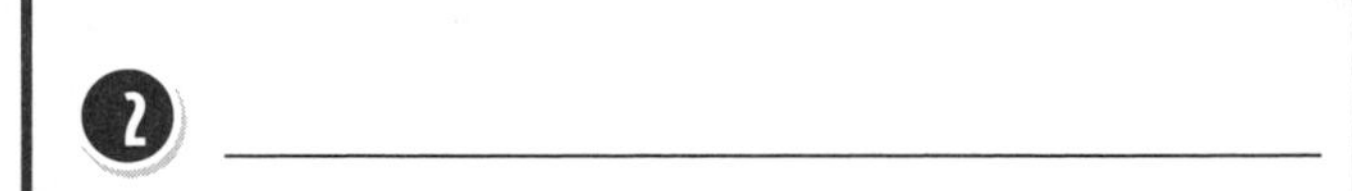

2. ______________________________

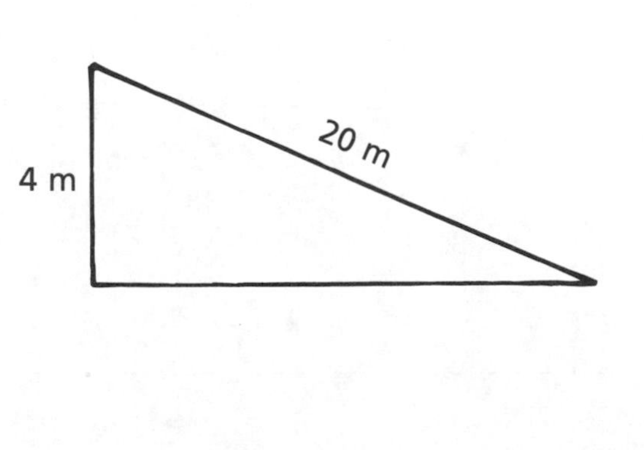

3. ______________________________

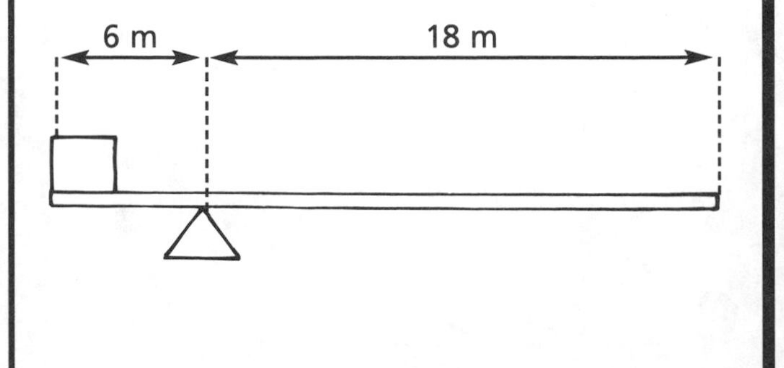

4. ______________________________

Name ______________________________ Date ______________

Pulleys

A **pulley** is a small wheel with a grooved rim that holds a rope or chain. Pulleys can be fixed or movable. Each pulley or wrap of the line allows you to trade distance for force. Use the terms in the word box to label the illustrations.

distance to move load	length of rope pulled	force	pulley

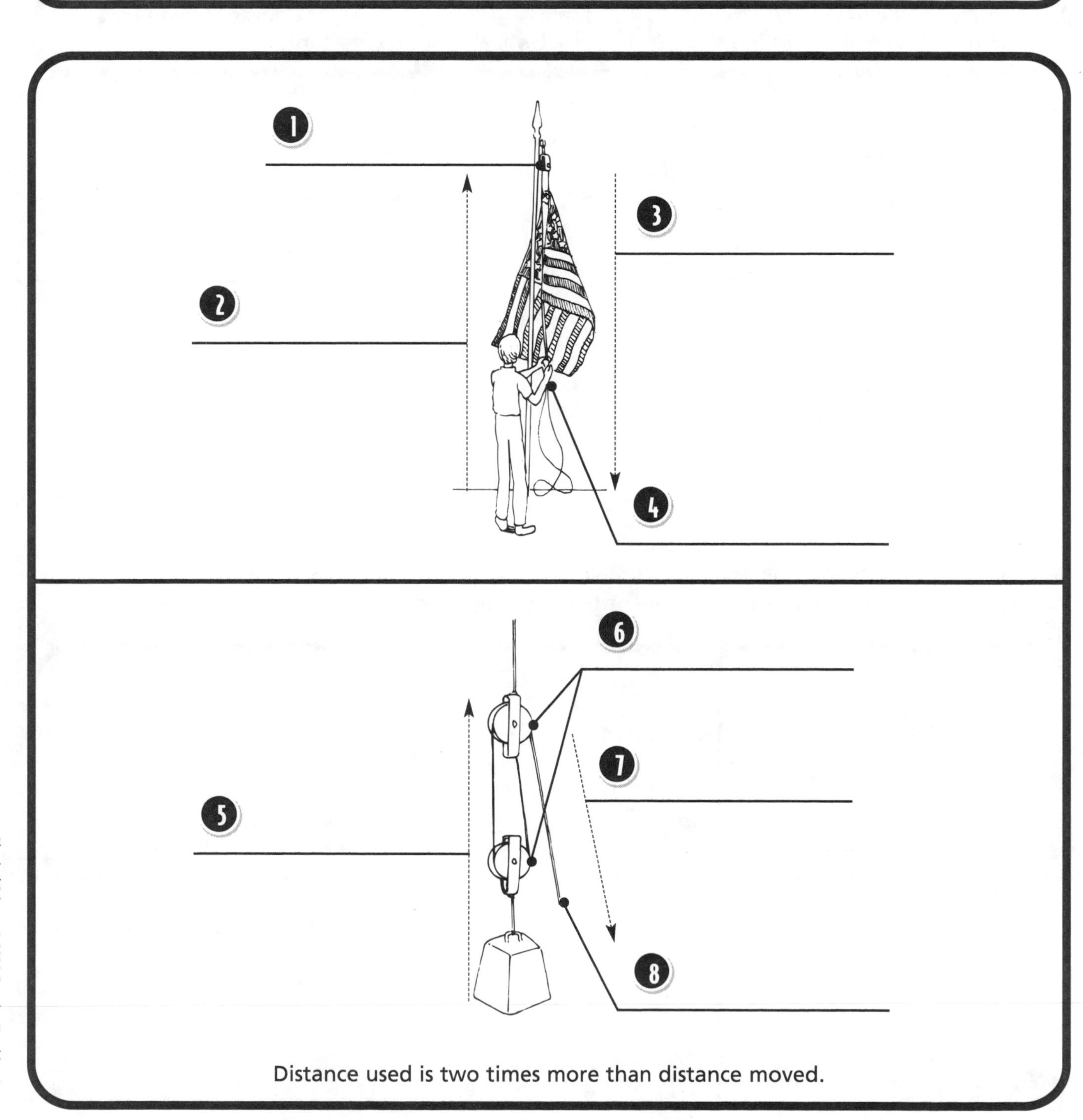

Distance used is two times more than distance moved.

Name ______________________________ Date ______________

Wheels and Axles

We see wheels on things every day. However, these kinds of wheels do not always do the same kind of work as the simple machine called the wheel and axle. The purpose of some wheels is to reduce friction by allowing an object to roll instead of drag. The simple machine is made up of a small wheel attached to a larger wheel. The small wheel is the axle. It is usually a rod that turns as a larger wheel turns. Use the terms in the word box to label the diagram.

distance axle uses	force	distance you use
large wheel makes turning easier	small wheel is difficult to turn	

1. ______________
2. ______________
3. ______________
4. ______________
5. ______________

Write **wheel and axle** if it operates like a simple machine to increase force.
Write **reduce friction** if that is the purpose of the object in the diagram.

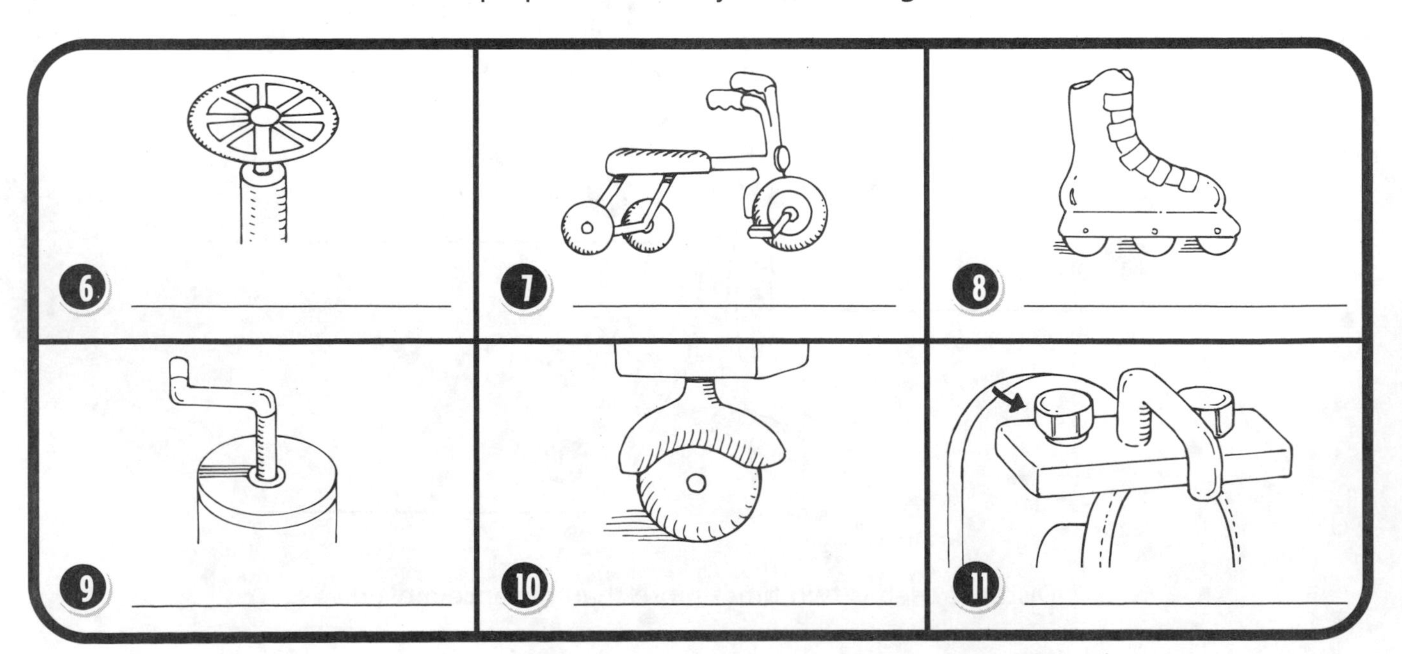

Name ______________________ Date ____________

Mechanical Advantage of Pulleys, Wheels, and Axles

Pulleys create mechanical advantage. Wheels and axles work together to create mechanical advantage. Gears are a type of wheel and axle. Use the formula to calculate the mechanical advantage for each diagram.

Mechanical advantage for pulleys: When the length of line you pull on is equal in length to the distance you move the load, the mechanical advantage (ma) is 1. Each time you add another length to the distance you must pull, by adding another wrap around a pulley, the mechanical advantage increases by 1 more.

Mechanical advantage for wheels and axles:

$$MA = \frac{\text{radius of the wheel}}{\text{radius of the axle}}$$

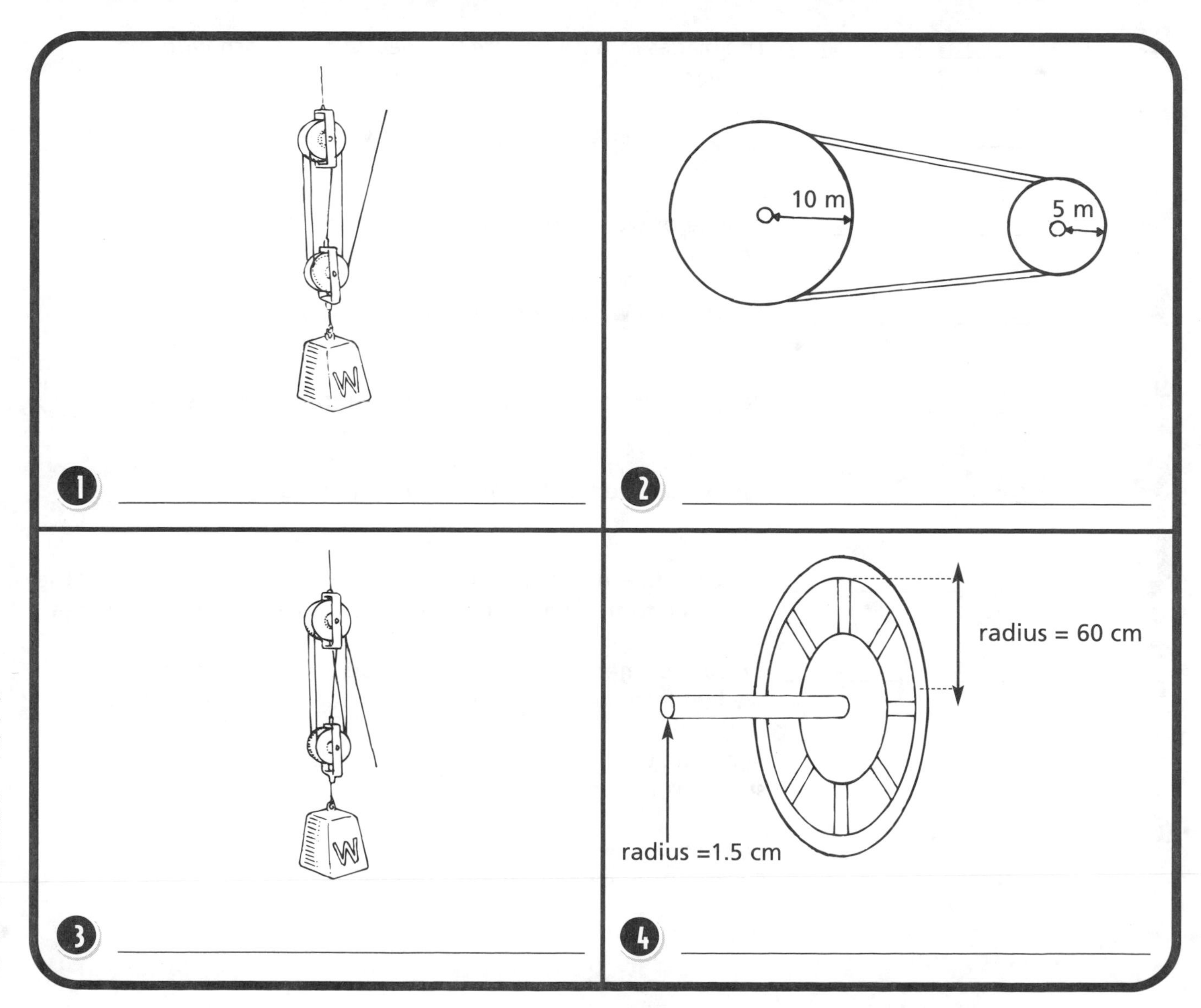

1 ______________________

2 ______________________

3 ______________________

4 ______________________

Name ______________________________ Date ____________

Wedges

A **wedge** is made up of two inclined planes. Unlike an inclined plane which does not move, the wedge moves to do work. Wedges can be used to lift things, separate things, or tighten and attach things in place. Use the terms in the word box to label each use of a wedge.

lift	separate	tighten	attach

1. ____________________ A nail is a wedge. Drive the nail into a block of wood.
2. ____________________ A doorstop is a wedge. Prop a door open with a doorstop.
3. ____________________ The blades of a pruner are wedges. Cut a branch with the pruners.
4. ____________________ The blades on a pair of scissors are wedges. Use the scissors to cut paper.
5. ____________________ A piece of wood can be a wedge. A man slips a wedge under a heavy box.
6. ____________________ A chisel is a wedge. An artist uses a chisel to carve a statue.
7. ____________________ Paper staples are wedges. Use staples to hold pages together.
8. ____________________ An axe is a wedge. Use an axe to split a log.
9. ____________________ Little wedges of wood can tighten a loose chair leg. Hammer the wedges into the space where the chair leg is attached to the seat.
10. ____________________ Wedges of metal can secure the head of an axe. Pound the wedges into the wood where the handle joins the axe head.
11. ____________________ The edge of a snow shovel is a wedge. Use the shovel to push snow off your driveway.
12. ____________________ The blade of a knife is a wedge. Be careful as you slice a piece of cheese for a hamburger with the wedge.

Name ______________________________ Date ______________

Screws

Screws are a simple machine that is a combination of other simple machines. Match each word in the word box to its description.

inclined plane	threads	turned
wheel and axle	handle	axle
wedge	force	distance
drill	jackscrew	

1. ______________ This is a variation of a screw that is designed to make holes in wood or other materials.
2. ______________ A screw is one of these wrapped around a cylinder.
3. ______________ With the use of a screwdriver, this is done to the screw as it enters the wood or other material.
4. ______________ When you are moving the screw with a screwdriver, you are really moving it with this simple machine.
5. ______________ Without the threads, a screw is basically a nail, which is an example of this.
6. ______________ When driving a nail, it takes a lot of this to push apart the wood as the nail enters it.
7. ______________ On a screw, the inclined plane forms these ridges in a spiral along the cylinder.
8. ______________ When turning a screw, it takes less force but you use more of this as the screw enters the wood.
9. ______________ This tool is a variation of a screw that allows a platform to raise and lower extremely heavy objects.
10. ______________ This part of the screwdriver acts as a wheel to do the work.
11. ______________ The shaft of the screwdriver and the screw itself act as one of these to do the work.

Name ______________________ Date ______________

Classifying Simple Machines

Classify the useful items listed in the word box by the main type of simple machine.

hammer	scissors	nut and bolt
staircase	step stool	bicycle gears
seesaw	crane	window blinds
front teeth	flagpole	bottle opener
fishing pole	door knob	faucet handle
hand drill	light socket	zipper
knife	ladder	ramp
wrench and pipe		

Levers	Screws	Inclined Planes
______	______	______
______	______	______
______	______	______
______	______	______
Wedges	**Pulleys**	**Wheel and Axles**
______	______	______
______	______	______
______	______	______
______	______	______

Name ______________________________ Date ______________

Uses for Simple Machines

Match the simple machines in the word box to the descriptions of the work that needs to be done. Some terms are used more than once

lever	pulley	wheel and axle
wedge	screw	inclined plane

1. ______________ A piano needs to be moved up to a third floor apartment.
2. ______________ Your sister wants to fasten a towel bar in the bathroom. You suggest one of these.
3. ______________ A large boulder is in your yard. You need to move it over about 3 feet.
4. ______________ You need to block a door to keep it open while you bring in groceries from the car.
5. ______________ Dad needs to take your older dog to the vet. The dog can no longer jump into the car on his own and he is too heavy to lift.
6. ______________ Every morning the flag needs to be raised to the top of the flagpole.
7. ______________ You remember to put the lid back on the jar of mayonnaise.
8. ______________ The trees have lost their leaves and you have a lot of them to collect.
9. ______________ Your little sister's treehouse is 6 feet off the ground and her friends need an easier way to get up to it.
10. ______________ You turn off the water to the sprinkler hose for your mom.
11. ______________ Your older brother chops some firewood because it's going to be a cold night.
12. ______________ While visiting an old farm, you need to draw a bucket of water up from the well.

Name ______________________ Date ______________

The Various Forms of Energy

Match each term in the word box to its definition.

sound energy	radiant energy	thermal energy
mechanical energy	nuclear energy	chemical energy
electrical energy	kinetic energy	potential energy
forms		

1. ______________ This is stored energy of an object or material. It is the energy that an object has due to its position.
2. ______________ This is the energy associated with movement of electrons through a wire or circuit.
3. ______________ This is the energy produced when an atom splits apart (fission) or when two atoms join to form one atom (fusion).
4. ______________ This is the energy of motion. The faster an object moves, the more of this it has.
5. ______________ This is the energy of an object due to the motion of its atoms and molecules. An object that feels hot has more of this inside it than it does after it has cooled down.
6. ______________ This is energy that can travel in waves and can move through empty space.
7. ______________ This is the energy stored in the connections between atoms. As chemical reactions take place to release these connections, this energy is released.
8. ______________ All energy can change from one of these to another.
9. ______________ This is the energy of vibrations carried through solids, liquids, or gases. It travels in waves, but it cannot move through empty space.
10. ______________ This is the energy of an object do to its motion, position, or condition. It is the combined total of potential energy and kinetic energy of an object.

Name ______________________________ Date ______________

Types of Energy

Energy can exist in many different forms. Each of these energy forms can be changed from one type to another and back again. Match each term in the word box to a description of each type of energy.

nuclear	sound	mechanical	radiant
thermal	electrical	chemical	

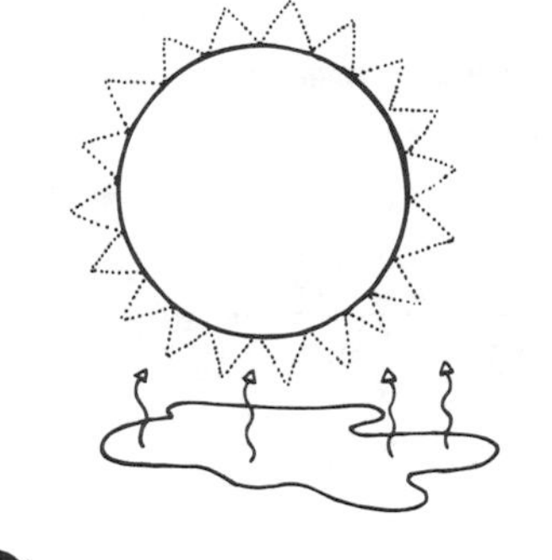

1. ______________

2. ______________

3. ______________

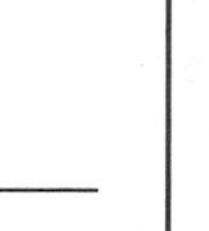

4. ______________

5. ______________

6. ______________

7. ______________

Name ______________________ Date ____________

Energy of One Kind or Another

Scientists classify all energy as kinetic energy or potential energy. At any given time, energy is being stored as potential energy or it is being given off as kinetic energy. The actions of a roller coaster can show how objects move from one type of energy to another. Use the terms in the word box to label the diagram.

electrical energy	potential energy changing to kinetic energy
mechanical energy	most potential energy
most kinetic energy	thermal energy

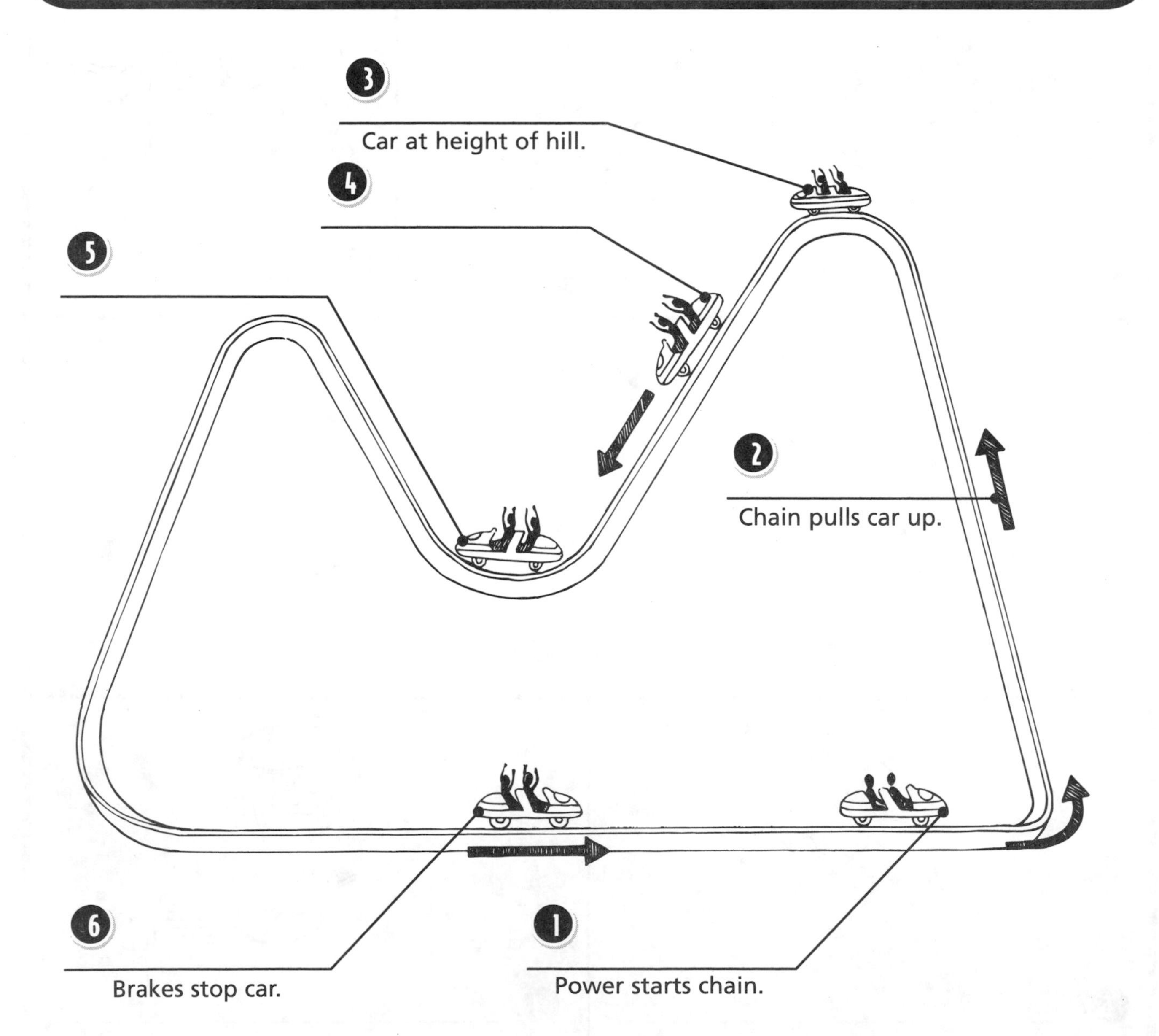

Name ______________________________ Date ______________

From One Form to Another

Energy can change form, but it cannot be created or destroyed. Under some conditions, such as those that produce nuclear energy, matter and energy can change into each other. However, the total amount of matter and energy does not change. Use the phrases in the word box to label the changes in energy forms.

electrical to sound	radiant to chemical	thermal to mechanical
chemical to thermal	mechanical to electrical	electrical to radiant
radiant to thermal	electrical to mechanical	

1. ______________ Plants use energy from the sun in the process of photosynthesis to manufacture glucose.

2. ______________ Solar panels placed on the top of a roof use the sun's energy to heat water for a swimming pool.

3. ______________ A lamp is plugged into a wall socket. Once the lamp is turned on, both light and heat come from the lamp.

4. ______________ The spinning turbine inside a hydroelectric plant generates electricity.

5. ______________ Some furnaces burn coal to heat buildings.

6. ______________ An alarm clock begins to beep and the radio turns on. The display shows it is 7:00 a.m.

7. ______________ Water is heated until it becomes steam. The pressure of the steam turns a turbine or generator.

8. ______________ Plug a fan into a wall socket. Turn on the fan and the blades begin to spin.

Name ______________________________ Date ______________

Energy Transfers

Energy can change from one form to another. We commonly change energy from one form to another as we use it. Match each term in the word box to its description. Some terms are used more than once.

radiant	chemical	electrical	thermal
mechanical	sound	nuclear	

1. ____________ When you shout at a friend, you are changing mechanical energy into this form.
2. ____________ When you set food under a heat lamp to warm it, you are changing radiant energy into this form.
3. ____________ When you turn on a CD player, you are using this form of energy to produce sound energy.
4. ____________ At a power plant nearby, matter is changed into energy to produce this form of energy.
5. ____________ Eating a healthy breakfast provides this form of energy, which is turned into mechanical and thermal energy as you play, study, and move.
6. ____________ Plug in a fan to a source of electrical energy. The electrical energy is transformed into this form to move the fan blades.
7. ____________ Plug a lamp into an electrical socket and you turn electrical energy into this form of energy.
8. ____________ Radiant energy from the sun is turned into this form when plants undergo photosynthesis.
9. ____________ As plants decompose, their chemical energy may become stored as coal or natural gas. These chemical energy sources can be turned into this form of energy as they are burned at a power plant.
10. ____________ As fuel is burned at a power plant, the energy released turns a turbine, whose movement is an example of this kind of energy.
11. ____________ Energy leaves a power plant in the form of this kind of energy.

Name ______________________________ Date ______________

Classifying Potential or Kinetic Energy

Various forms of energy can be classified as being either a potential energy source or a kinetic energy source. Classify the phrases in the word box as examples of potential or kinetic energy.

standing at the top of a slide
wind up for the pitch
juice in an orange
move downhill in a roller coaster
roll down a grassy hill
an unburned lump of coal
throw a curve ball
a battery
frog leaping into the water
book falls from a high shelf
move down a slide
frog sitting on a lily pad
book on a high shelf
a speeding car
execute a swan dive
a parked car

Potential Energy	Kinetic Energy

Name ______________________ Date ____________

An Electric Generator

Use the terms in the word box to label the diagram of an electric generator.

wheel and axle	magnet	drive shaft
wire coil	source of mechanical energy	

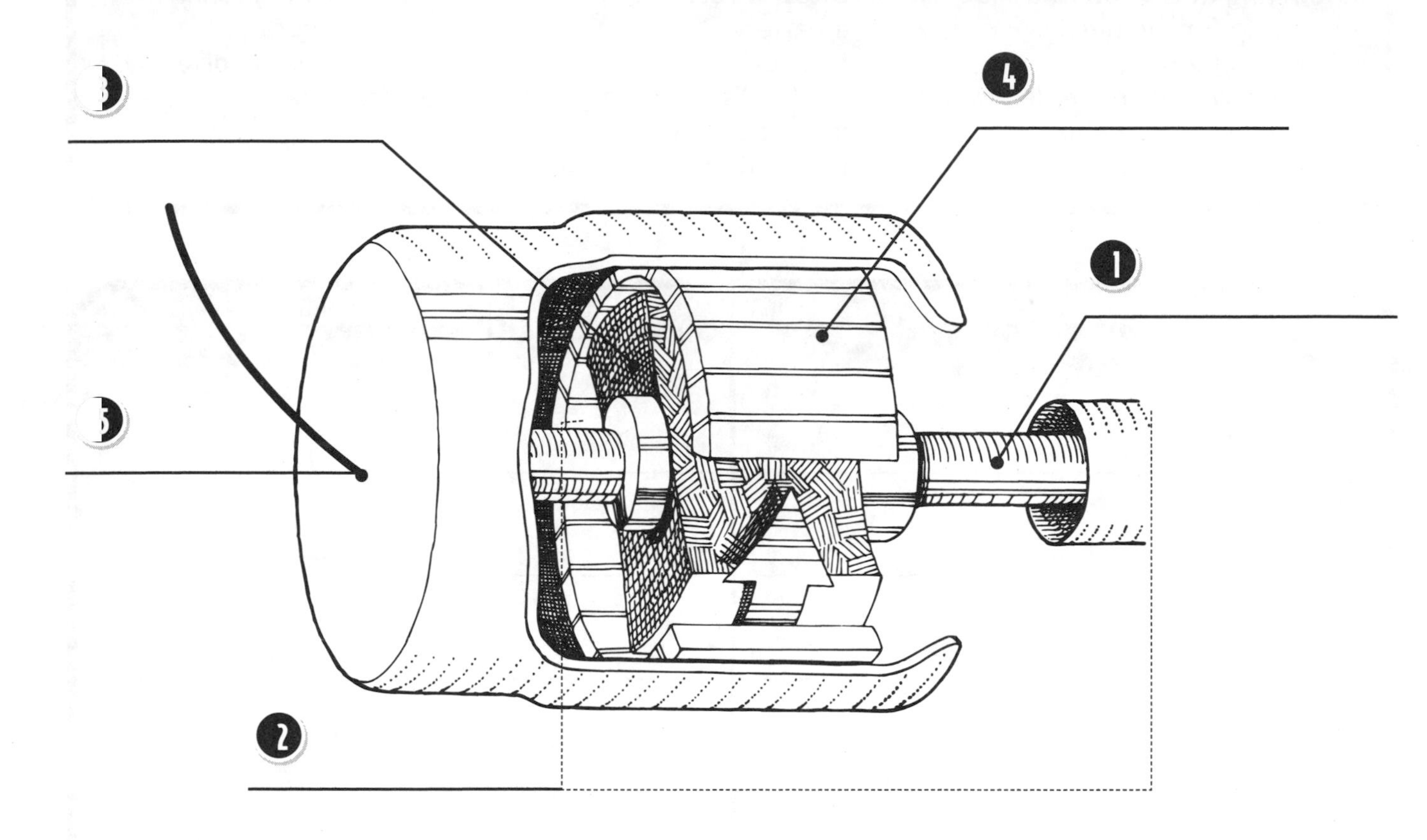

Match each term in the word box above to its definition.

6. ______________ The magnet and the drive shaft in a generator are an example of this kind of simple machine.

7. ______________ Connected to the magnet, this turns the magnet inside a coil of wire.

8. ______________ The moving magnet makes electrons flow within this, generating an electric current.

9. ______________ Having two opposite poles, its force due to its spinning is used to move electrons.

10. ______________ Used to turn the drive shaft, it can come from moving water or heating water to create steam.

Name ______________________________ Date ______________

Energy from a Nuclear Reactor

Nuclear reactors are basically heat engines. Nuclear fission releases energy from the fuel. This process releases energy, much of it in the form of heat, which can then be used to do work. Match each term in the word box to its definition.

reaction chamber	control rods	moderator
coolant	turbine generator	shield
fuel	transformer	heat exchanger

1. ______________ This protects the surrounding environment, human beings, and the reactor itself from radiation.
2. ______________ This changes the voltage of electricity produced.
3. ______________ This is used to absorb thermal energy released by the reaction. It keeps the reactor temperature at reasonable levels.
4. ______________ These control the rate of the nuclear reaction by absorbing neutrons.
5. ______________ This area of the reactor contains the fuel and provides a site for reactions to occur.
6. ______________ Powered by the steam, its rotating blades turn the shaft, which in turn generates electrical energy.
7. ______________ This component transfers heat from the reaction chamber.
8. ______________ This is fissionable material that is used to form nuclear energy. It is usually from a uranium or plutonium source.
9. ______________ This material slows down neutrons to allow them to be captured by the nuclei of the fuel.

Name ______________________________ Date ______________

Parts of a Nuclear Reactor

Use the terms in the word box to label the diagram of a nuclear reactor.

reaction chamber	control rods	moderator	coolant
turbine generator	shield	fuel	transformer
heat exchanger			

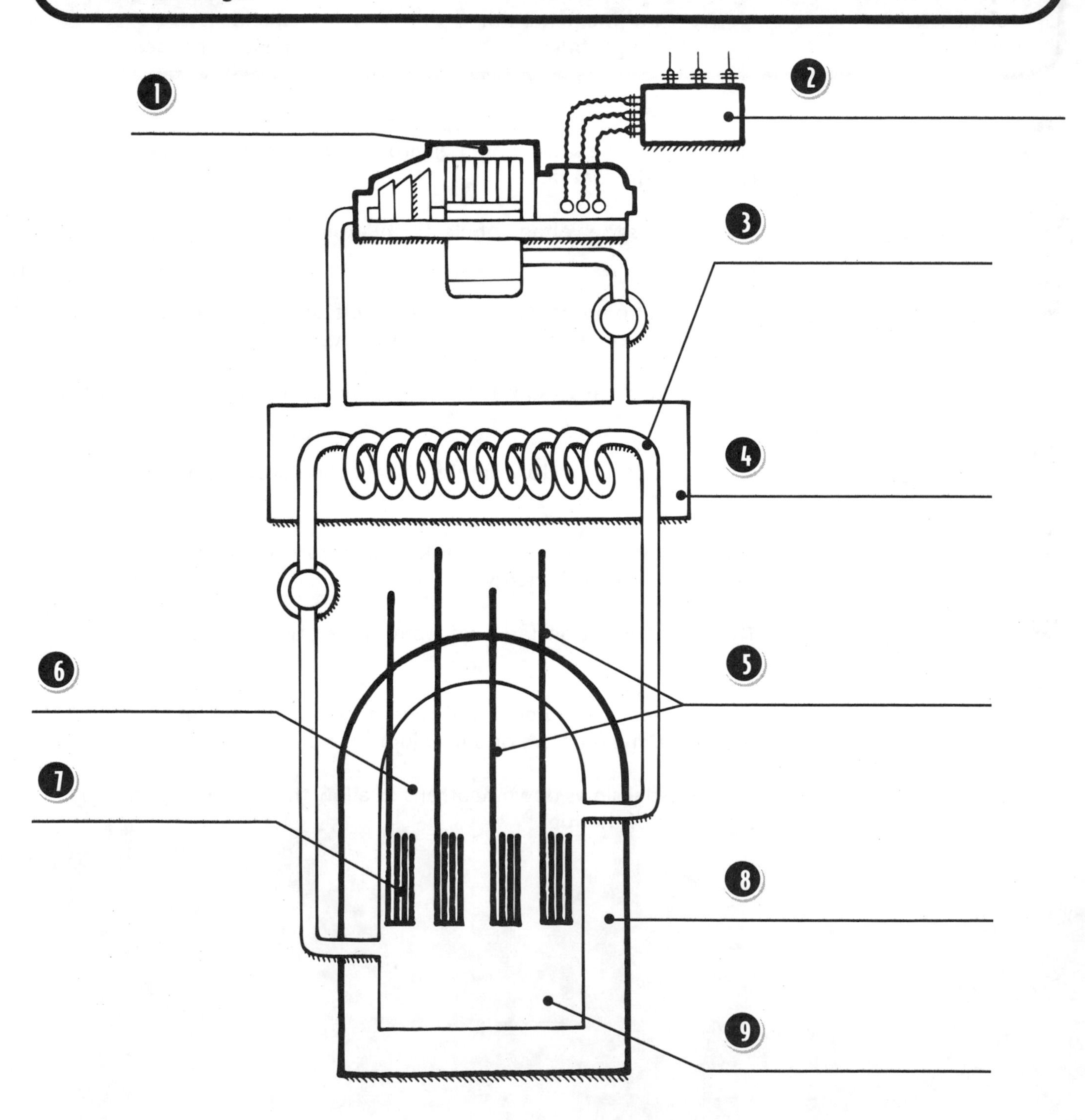

Name ______________________________ Date ______________

Alternative Energy Sources

Alternative energy refers to energy sources that are not based on the burning of fossil fuels or the splitting of atoms. Fossil fuels are nonrenewable and create pollution when burned. Nuclear energy produces harmful radioactive waste products. Alternative sources can provide clean renewable energy for use in homes and communities. Match each term in the word box to its description.

cells	windmill	hydroelectric	geothermal
tidal	solar energy	efficient	sun
wind technology	wind	hotspots	

1. ______________ The most renewable energy resource, it provides Earth with 35,000 times more energy than human activity uses altogether.
2. ______________ The apparatus used to capture wind power and turn it into electrical energy.
3. ______________ An indirect source of solar energy, it is caused by the heating and cooling of Earth's surface.
4. ______________ This energy is obtained from Earth's internal heat and can be used to generate steam to run a steam turbine.
5. ______________ This is an area of reduced thickness in Earth's mantle that allows heat from the interior to reach the outer crust. Volcanoes, vents, and geysers are examples.
6. ______________ One of our earliest energy resources, this technology is dependent upon weather and location.
7. ______________ This is the origin for radiant energy and most other types of energy in our solar system.
8. ______________ This form of energy is generated by the effect of gravity between Earth and the moon.
9. ______________ This energy source comes from rivers by utilizing the potential energy stored in the moving water.
10. ______________ These are units that convert sunlight directly into electricity.
11. ______________ This describes resources that produce energy effectively with a minimum of waste, expense, or unnecessary effort.

Name ______________________________ Date ______________

Terms of Light

Light is a form of energy that originates at the sun. It is a form of energy that is visible to the human eye. Match each term in the word box to its definition.

visible light spectrum	crest	trough
hertz	wave velocity	frequency
reflection	wavelength	prism
refraction	photon	light

1. ____________ This refers to the bending of light waves when they pass through another substance.
2. ____________ This is the bounce of a light wave off another object.
3. ____________ This is a particle of light.
4. ____________ This term refers to the highest point of a wave.
5. ____________ This is a type of electromagnetic radiation.
6. ____________ This is the distance between corresponding points on two waves.
7. ____________ This is a continuous band of colors arranged according to wavelength or frequency.
8. ____________ This is the lowest point of a light wave.
9. ____________ This is triangular-shaped glass or other transparent material that refracts white light into a spectrum of colors.
10. ____________ This is calculated by multiplying frequency times wavelength.
11. ____________ This is a measurement unit for frequency.
12. ____________ This refers to the number of waves that pass a given point in one second.

Name ________________________ Date ____________

Diagram of a Wave

Both light and sound travel in waves. Use the terms in the word box to label the parts of a wave.

amplitude	wavelength	crest	equilibrium	trough

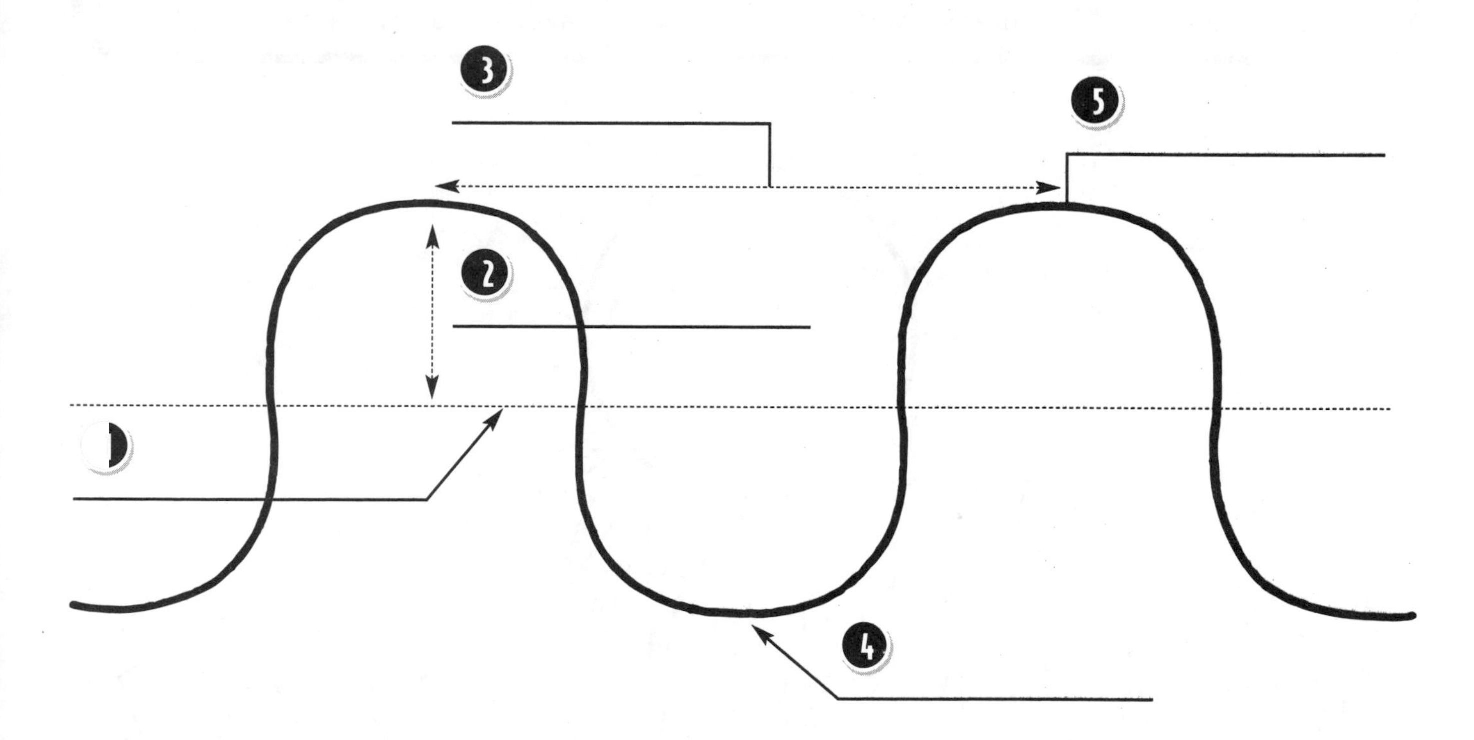

Match each term in the word box above to its description.

6. ____________ This is a measure from a point on one wave to the corresponding point on the next wave.
7. ____________ This is the lowest point on a wave.
8. ____________ This is the highest point on a wave.
9. ____________ This is the distance a wave rises or falls from its equilibrium.
10. ____________ This describes when the wave is at a rest position.

Name ____________________ Date ____________

Light Waves

Light is an example of radiant energy. The human eye is only able to see light of a certain wavelength. However, there are other wavelengths that are not visible to the human eye. Use the terms in the word box to label the diagram of the electromagnetic energy spectrum.

long radio waves	X-rays	microwaves	ultraviolet rays
visible light	infrared rays	short radio waves	gamma rays

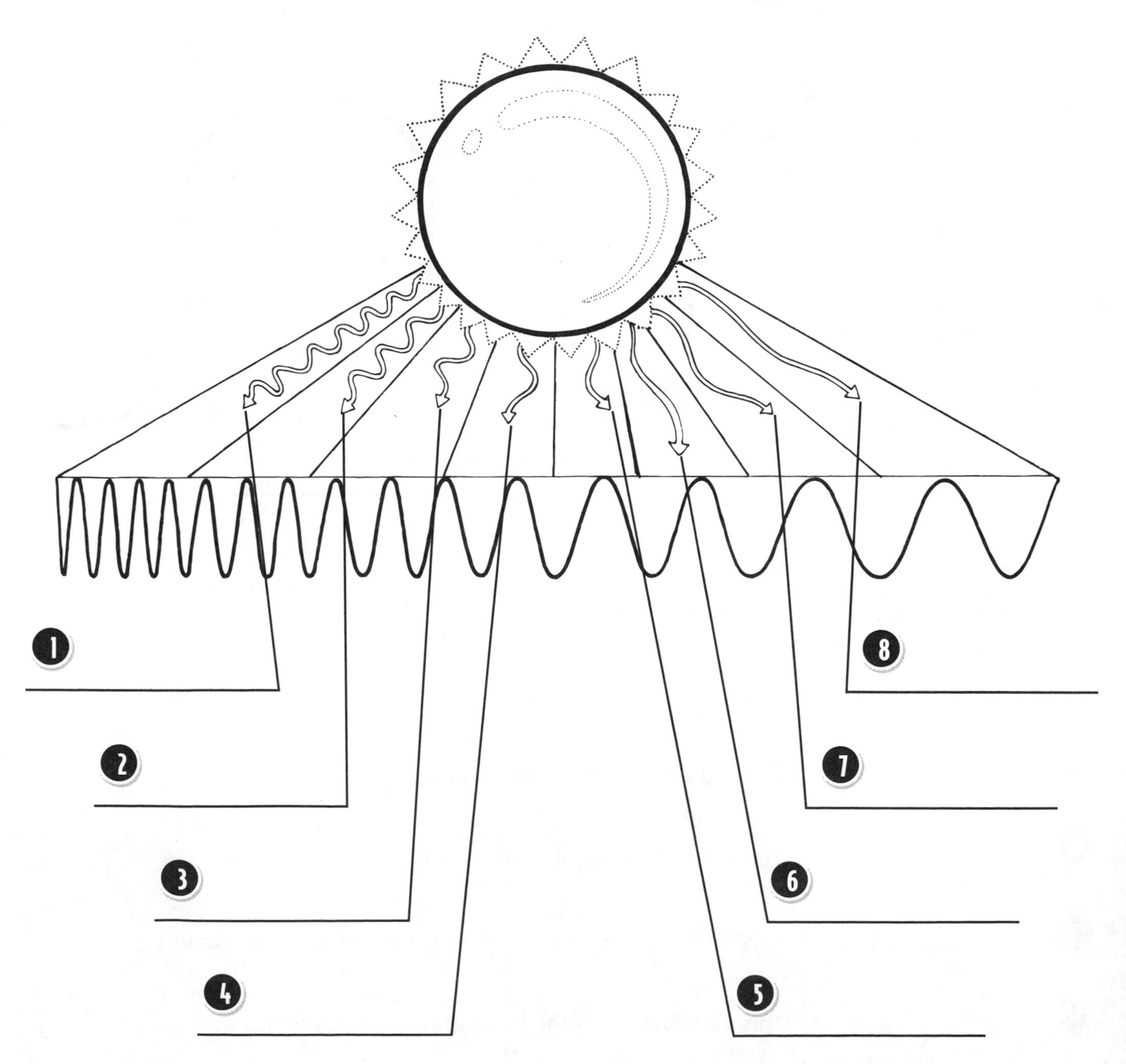

Name ______________________________ Date ______________

Uses of Electromagnetic Energy

Different types of electromagnetic energy have different uses and effects. Classify the phrases in the word box under each type of light wavelength.

kills organisms that spoil food	treats some cancers	shows breaks in bones
shows heat loss in buildings	shows cavities in teeth	used to kill germs
allows us to see	creates a tan	creates a rainbow
television signals	used to cook food	radio signals
portions of phone calls	radar	sun's heat
fire's heat	cell phone signals	causes a sunburn
maritime communication		

Gamma Rays	X-Rays	Ultraviolet Rays
		______________
______________	______________	______________
______________	______________	______________

Visible Light	Infrared Rays	Microwaves
	______________	______________
______________	______________	______________
______________	______________	______________

Radio Waves	
______________	______________
______________	______________

Name ______________________________ Date ______________

White Light Spectrum

Our eyes are sensitive to only a very small part of the electromagnetic spectrum. This portion that we can see is called **visible light**. This white light is a combination of the colors of the light spectrum. Each color is a different wavelength. Violet is the color with the shortest wavelength. Red is the color with the longest wavelength. Use the terms in the word box to label the colors of the spectrum and their wavelengths.

red	blue	violet	green
indigo	yellow	orange	492 – 475 nm
445 – 390 nm	780 – 622 nm	622 – 597 nm	577 – 492 nm
475 – 445 nm	597 – 577 nm		

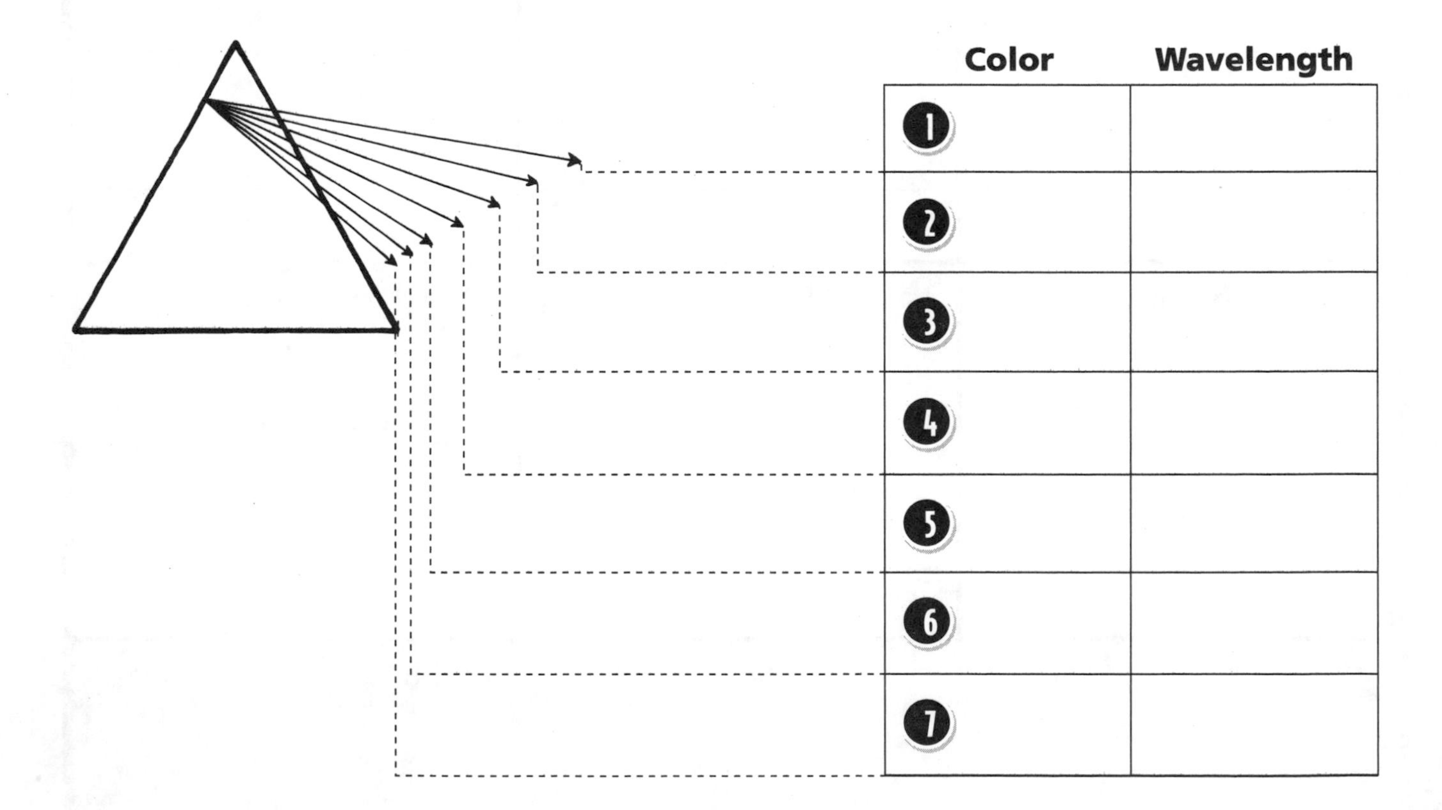

	Color	Wavelength
1		
2		
3		
4		
5		
6		
7		

Name ______________________________ Date ______________

Light Rays and Concave Lenses

Concave lenses work to make something look smaller. Eyeglasses have either a convex or concave surface. The lenses can bend the light just the right amount to focus an image. Use the terms in the word box to label the diagrams. Then draw the path of light through the lenses to show how they help people see. Use each term once for each set of images.

focal point	image	object	lens

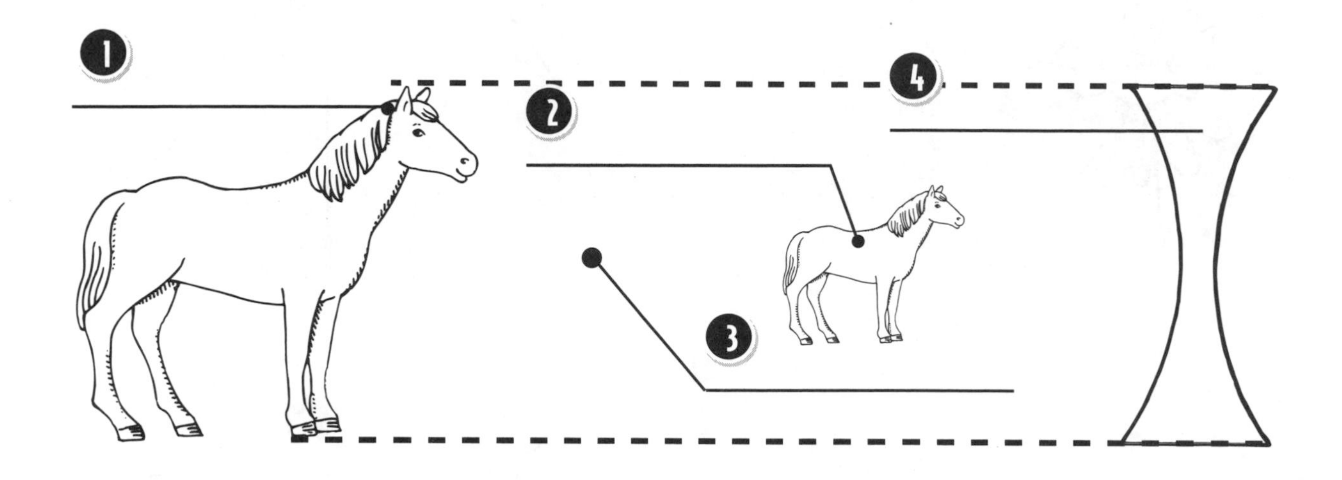

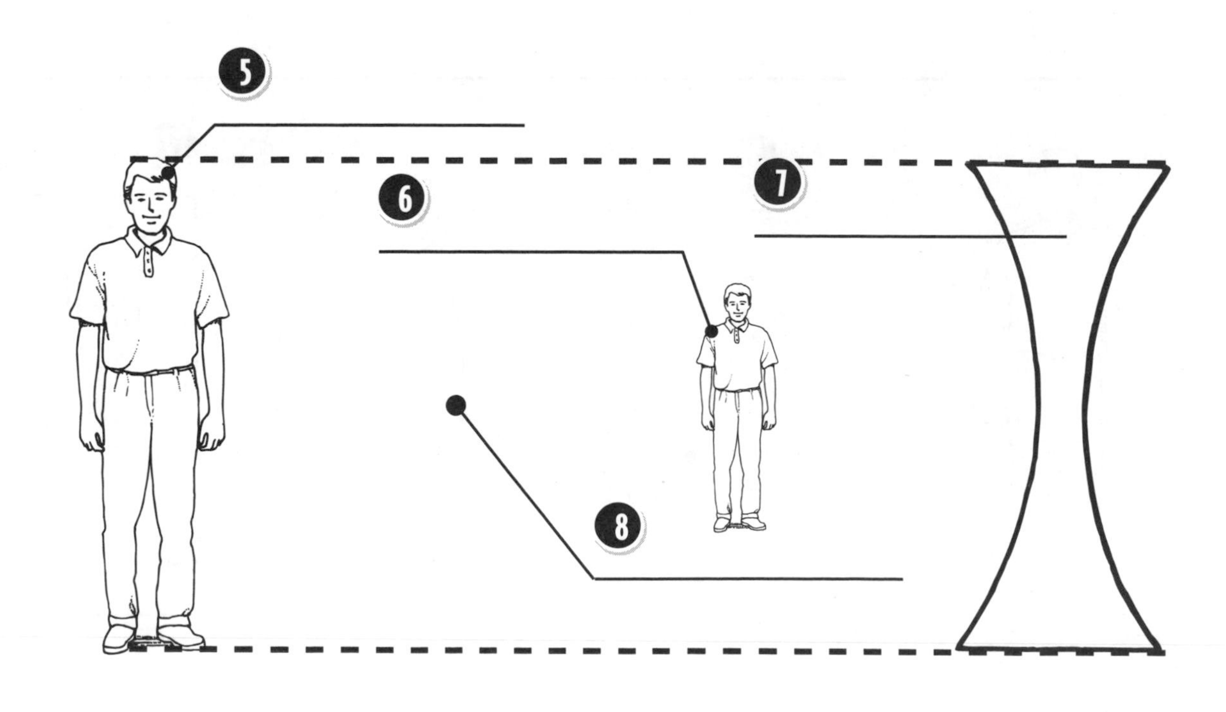

Name ______________________ Date ____________

Light Rays and Convex Lenses

Convex lenses work to make something look larger. Convex lenses are used in binoculars, telescopes, and magnifying glasses. Use the terms in the word box to label the diagrams. Draw the path of light through the lenses to show how they help people see.

object	focal point	inverted image	lens

1 ____________

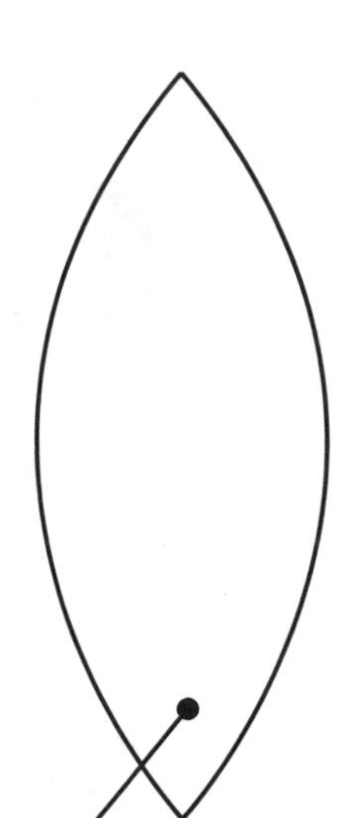

2 ____________

3 ____________

4 ____________

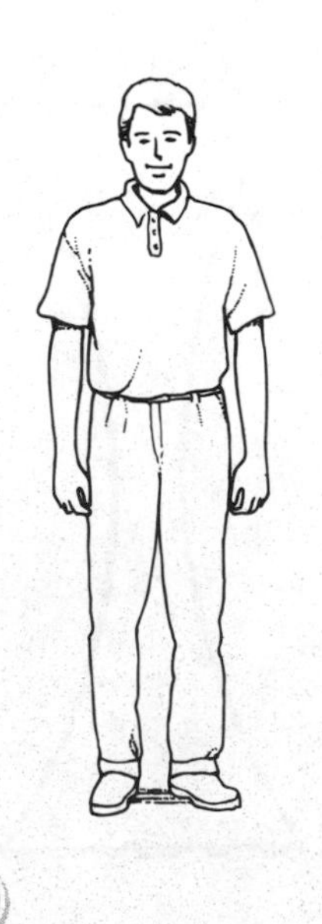

5 ____________

6 ____________

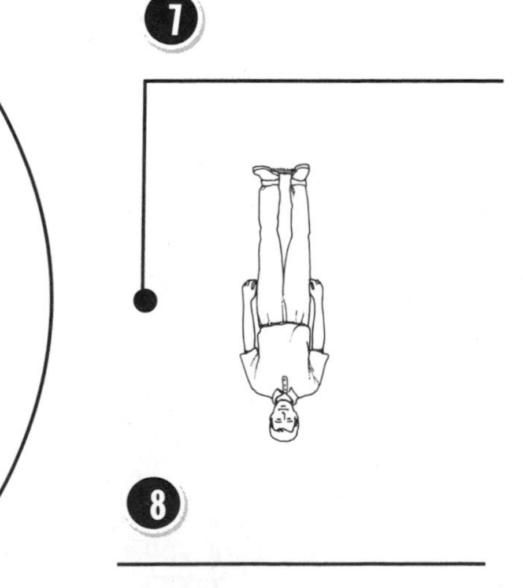

7 ____________

8 ____________

Name ______________________________ Date ______________

Reflection

A beam of light strikes a reflecting surface. The angle between a line perpendicular to the surface and the beam as it comes in contact with the surface is called the **angle of incidence**. The **angle of reflection** is how the beam bounces off the surface as measured from a line perpendicular to the point of reflection. The angle of incidence and the angle of reflection are equal. Draw the expected path of the light rays as they reflect off these plane mirrors.

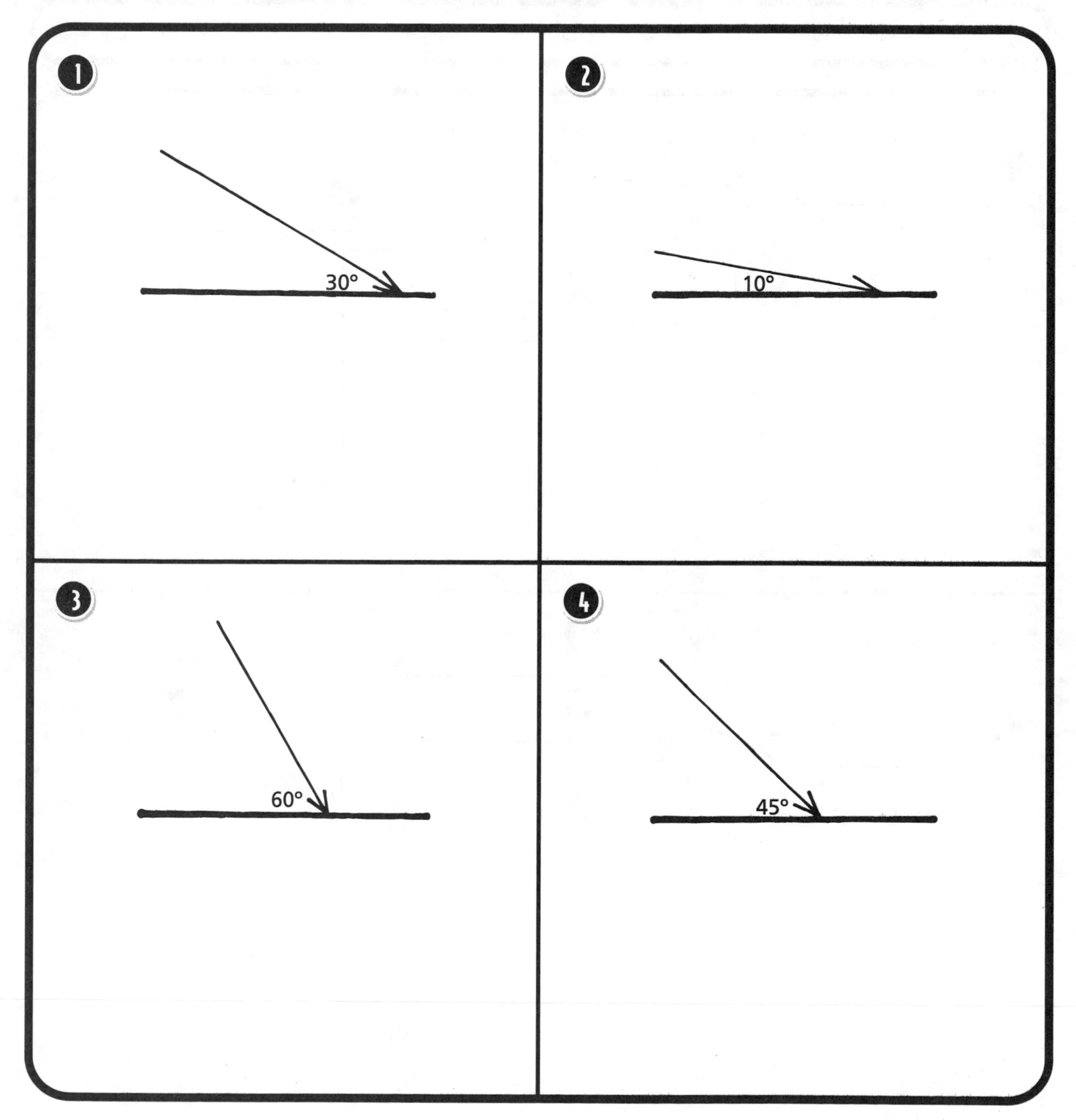

Name ______________________ Date ____________

Refraction

Refraction of light occurs when a light ray changes mediums. Light traveling from air and going into water is one example. The speed of the light ray changes as it enters another medium. In almost every case, the direction of the light ray changes also. Use a ruler to draw the path of the light ray as it passes through each substance in the diagrams. Then use the terms in the word box to label the diagrams.

refractive angle	normal	light ray	incident angle	surface

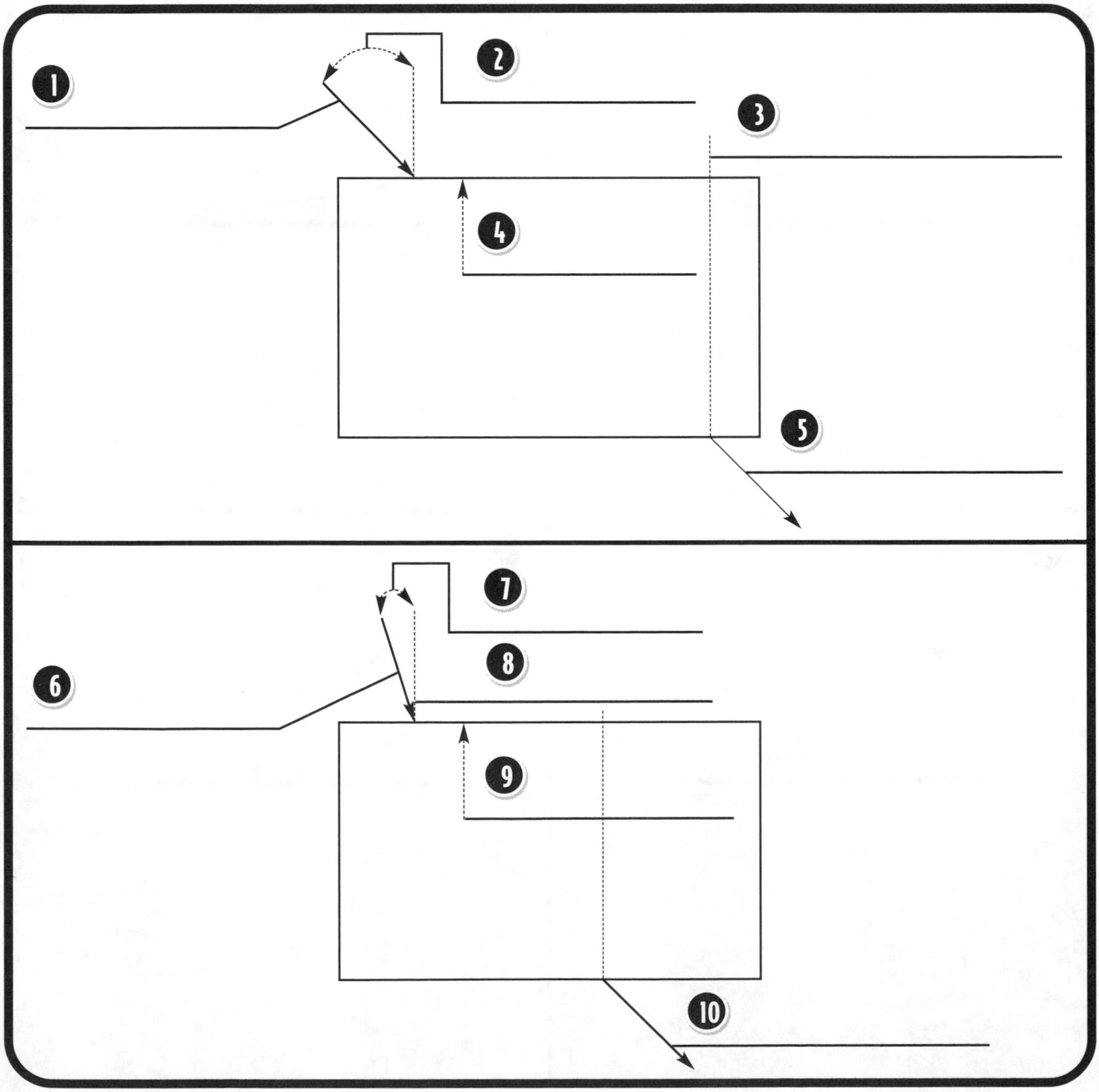

1 ______________________
2 ______________________
3 ______________________
4 ______________________
5 ______________________
6 ______________________
7 ______________________
8 ______________________
9 ______________________
10 ______________________

Name ______________________ Date ____________

Classify Refraction Materials

Visible light can pass through some types of matter but not others. Light can pass through some materials almost unchanged. Other materials scatter light or do not allow it to pass through at all. Use the first set of terms to label each material's effect on light. Then classify the second set of terms into the correct columns.

light passes through	light does not pass through	light scatters
aluminum foil	plastic wrap	cardboard
window glass	plywood	textured glass
wax paper	construction paper	oiled paper
velvet	tracing paper	clear water
muddy water	mud	prism
sheer fabric	car windshield	

Transparent	Translucent	Opaque
Effect: ____________	Effect: ____________	Effect: ____________
Examples	**Examples**	**Examples**
____________	____________	____________
____________	____________	____________
____________	____________	____________
____________	____________	____________
____________	____________	____________
____________	____________	____________

Name ______________________________ Date ______________

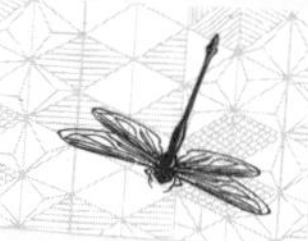

Diagram of a Sound Wave

A **sound wave** is an example of a pressure wave. A **pressure wave** is created by vibrating objects and moves through a medium from one location to another through alternating areas of compression and rarefaction. Use the terms in the word box to label the parts of a sound wave.

sound source	wavelength	compression
rarefaction	equilibrium	

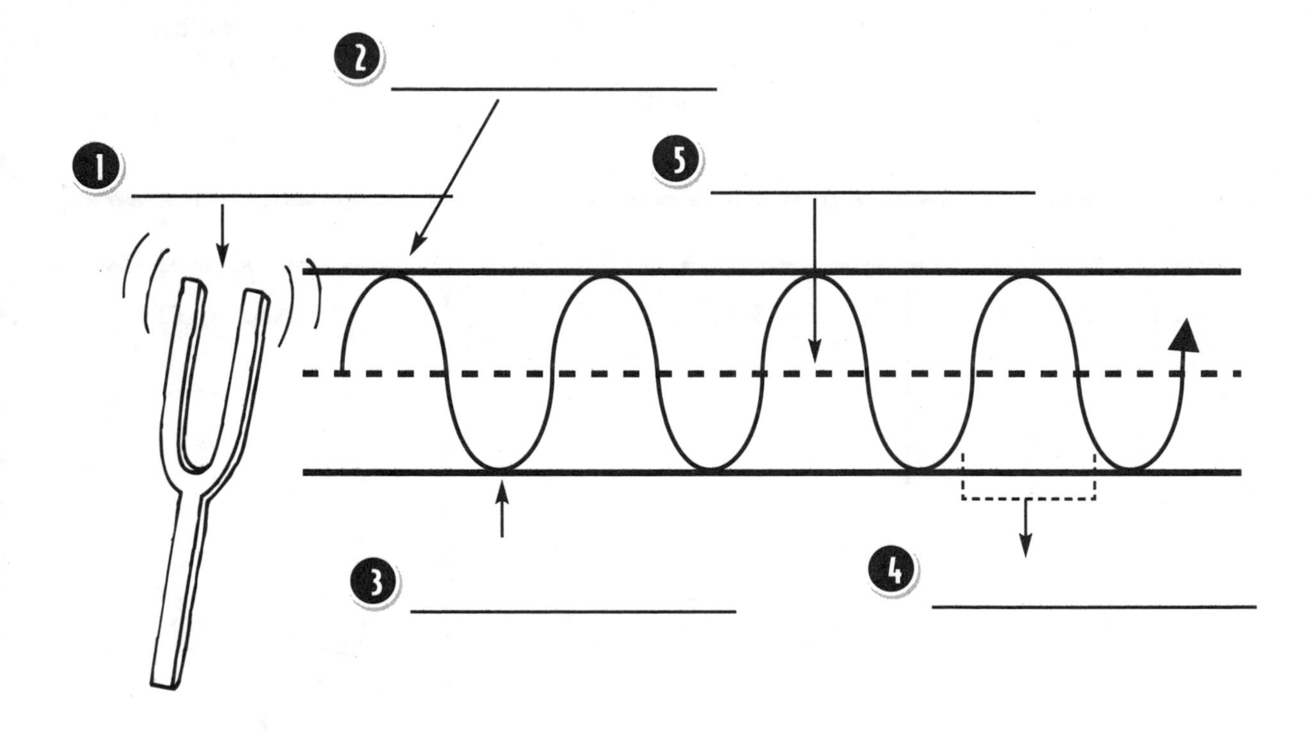

Match each term in the word box above to its description.

6. ______________ This describes where the pressure returns to its rest position.

7. ______________ This describes the part of a pressure wave where the vibrating particles are far apart.

8. ______________ This is the origin of the sound vibration.

9. ______________ This describes the part of a pressure wave where the vibrating particles are close together.

10. ______________ This is measured from a point on one wave to the corresponding point on the next wave.

Name ______________________ Date ____________

How Do These Sound?

Match each term in the word box to its definition.

vacuum	decibel	compression	ultrasonic
octave	pitch	Doppler	acoustics
amplitude	harmony	rarefaction	loudness

1. ______________ This is a unit that measures the intensity or loudness of sound.

2. ______________ This is the science of sound dealing with the production, effects, and transmission of sound waves through various mediums.

3. ______________ This is another word for the volume of sound. This increases as the amplitude of sound waves increases.

4. ______________ This refers to the range of eight notes on a musical scale.

5. ______________ This is the absence of matter such as air.

6. ______________ This describes sound vibrations above what is audible by the human ear.

7. ______________ The maximum value of a sound wave, measured from its equilibrium.

8. ______________ This effect is the apparent change in wavelength of sound or light caused by the motion of the source, the observer, or both.

9. ______________ This property of sound varies with the frequency of vibrations.

10. ______________ This describes the positive portion of a pressure wave by which vibrating particles are closer together.

11. ______________ This describes the negative portion of a pressure wave in which the vibrating particles are farther apart.

12. ______________ This describes a sound that is pleasing to the ear.

Name ______________________________ Date ______________

Sound on a Frequency Scale

Of all the sounds that are made, humans can hear only a small range. All musical instruments make sounds within that range. Many animals are capable of hearing or making sounds that are both higher and lower than humans. Match the words in the word box to each description.

ultrasonic	decibel	speed of sound	longitudinal
pitch	intensity	vibrates	compression
amplitude	Doppler effect	rarefaction	fluctuations

1. ______________ This varies but at sea level it measures 740 miles per hour (1190 kilometers per hour).
2. ______________ A sound wave is this kind of wave.
3. ______________ This refers to the height of a sound wave.
4. ______________ This phenomenon can be used to determine an object's speed and direction. It is the apparent change in the length of a sound wave due to the object's or observer's position.
5. ______________ This refers to the frequency of a sound vibration.
6. ______________ An object produces sound when it does this in matter.
7. ______________ This unit of measurement is used to express the intensity of a sound.
8. ______________ Sounds of frequencies above the range of normal human hearing are called this.
9. ______________ This occurs when a vibrating object causes air molecules to collide with those next to them.
10. ______________ As a vibrating object pulls away from the surrounding air molecules it causes a drop in pressure called this.
11. ______________ In order to hear a sound, your ear must sense these changes in air pressure.
12. ______________ This refers to the magnitude of a sound.

Name ____________________ Date __________

Wave Velocity

Wave velocities tell the direction and speed of a wave. Wave velocities are measured in meters per second. Use the formula in the example box to calculate the wave velocities.

Velocity = wavelength x frequency

1. A wave has a frequency of 2.5 hertz. The wavelength of the sound produced is 3.4 meters. What is the velocity of the wave?

   __________ meters per second

2. A tuning fork has a frequency of 280 hertz. The wavelength of the sound produced is 1.5 meters. What is the velocity of the wave?

   __________ meters per second

3. A wave has a frequency of 135 hertz. The wavelength of the sound produced is 5.7 meters. What is the velocity of the wave?

   __________ meters per second

4. A tuning fork has a frequency of 220 hertz. The wavelength of the sound produced is 4.2 meters. What is the velocity of the wave?

   __________ meters per second

Name ______________________________ Date ______________

How Loud Is Too Loud?

Sound is measured in decibels, which describes the intensity of sound. Each increase of 10 decibels means that the intensity of sound increases by a factor of 10. Some of the sounds we are exposed to can damage our hearing. Match the phrases in the word box to each description. Some phrases are used more than once. Then write **hearing damage** for those decibel levels considered harmful to hearing.

96 decibels	140 decibels	10 decibels	20 decibels
110 decibels	80 decibels	40 decibels	70 decibels
60 decibels	160 decibels	150 decibels	94 decibels
90 decibels			

1. ______________ A jet engine at take-off ______________
2. ______________ Busy city traffic during rush hour ______________
3. ______________ Inside a public library ______________
4. ______________ Normal human breathing ______________
5. ______________ A person whispering______________
6. ______________ Normal human speech______________
7. ______________ A lawnmower three feet away ______________
8. ______________ A rock concert ______________
9. ______________ A ringing alarm clock ______________
10. ______________ A power saw three feet away ______________
11. ______________ A blast on a trumpet ______________
12. ______________ Inside a car______________
13. ______________ Inside a subway ______________
14. ______________ A food blender three feet away______________

Name ______________________ Date ______________

Electrical Terms

Electricity is energy made available by the flow of an electric charge through a conductor. Electricity is possible because of the behavior of negative and positive charges (electrons and protons) and their attraction and repulsion. Electricity can be static or current. Match each term in the word box to its definition.

electric current	lightning	circuit	volt
kilowatt	static electricity	watt	direct
resistance	conductivity	ammeter	alternating

1. ______________ This is a sudden, visible discharge of electricity from one cloud to another or from a cloud to the ground.
2. ______________ This is a unit of electrical pressure. It is used to measure how strongly the electrons in a wire are pushed.
3. ______________ This is a material's opposition to the flow of electric current; measured in ohms.
4. ______________ This is an electric current that periodically reverses its direction.
5. ______________ This refers to a continuous flow of electric charge through a material that conducts electricity.
6. ______________ This is the ability of a material to allow the flow of electrical current.
7. ______________ This is a path through which electrical current flows.
8. ______________ This is a unit of electrical power. It is used to measure how fast electrical energy is used.
9. ______________ This is an instrument for measuring electric current in amperes.
10. ______________ This refers to an electric current that flows in one direction and does not reverse direction.
11. ______________ This is the electrical charge that builds up due to friction between two dissimilar materials.
12. ______________ A measurement equal to 1000 watts, abbreviated as *kW*.

Name ______________________________ Date ______________

Magnetic Terms

Magnetism is a property, possessed by certain materials, to attract or repel similar materials. Magnetism is associated with moving electricity. Match each term in the word box to its definition.

pole	magnet	magnetic field	repel	attract
compass	lodestone	magnetism	electromagnet	Gauss meter

1. ____________ This is a rock that possesses magnetic properties and attracts iron. It is also known as magnetite.
2. ____________ This means to drive away or force backwards.
3. ____________ This is an instrument for determining direction. It has a magnetic needle that rotates freely until it points to the magnetic north.
4. ____________ This is a type of magnet. An iron or steel core is magnetized by the electric current in the coil of insulated wire wound around it.
5. ____________ This is an object that attracts iron and some other materials by virtue of a natural or induced force field surrounding it.
6. ____________ This means to cause to come near, as by some special quality or action.
7. ____________ This describes the force itself that attracts, iron, nickel, and cobalt.
8. ____________ This is the name for one of the two ends of a magnet where the magnetism seems to be concentrated. It also refers to either end of the axis on with Earth spins.
9. ____________ This refers to a force field generated by moving electrical charges.
10. ____________ This is an instrument that detects and measures magnetic fields.

Name ______________________________ Date ______________

Types of Magnets

There are three main types of magnets. **Temporary magnets** are made from materials that are easily magnetized, such as an iron needle, but gradually lose magnetism once the magnetic field is removed. **Electromagnets** are very strong and created by placing a metal core inside a coil of wire that carries an electrical current. The moving current magnetizes the metal. **Permanent magnets** are those that, once magnetized, do not easily lose their magnetism. These are the magnets with which we are most familiar. Use the terms in the word box to identify the examples of magnets. Some terms are used more than once.

horseshoe magnet	cylindrical magnet	disk magnet	ring magnet
cow magnet	bar magnet	U-shaped magnet	electromagnet

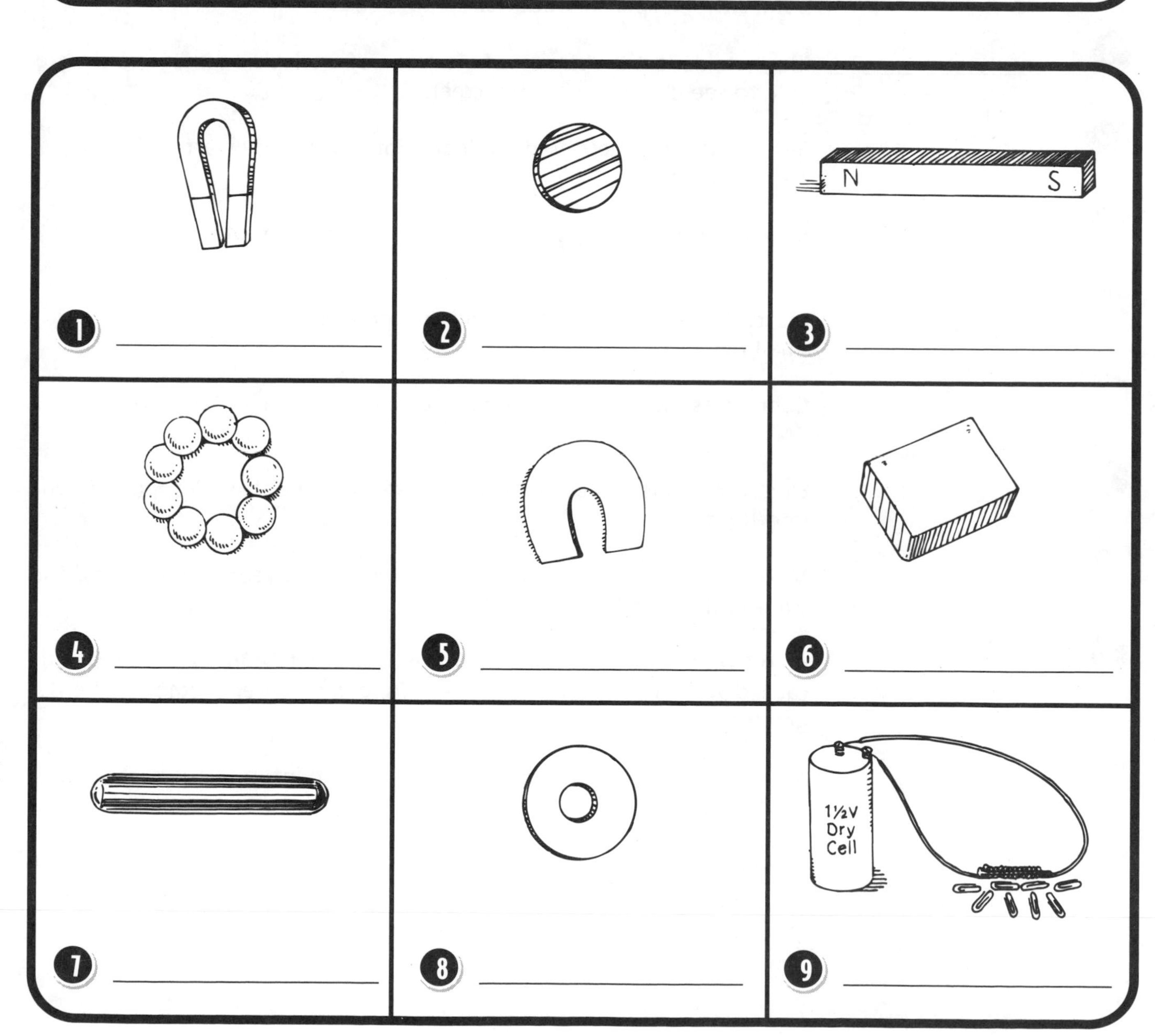

Name ______________________________ Date ______________

How Magnets Are Used

The properties of magnets that cause them to attract or repel other objects have many applications in our everyday lives. Use the phrases in the word box to classify each of the magnet uses.

used as separators	hold or attract objects	used for detection	use magnetic repulsion

1. ______________ Utility workers use magnets to locate underground pipes.
2. ______________ Vending machines use magnets to separate slugs from actual coins before they dispense products.
3. ______________ Magnets deflect the stream of electrons of a television image so we are able to see the image on the screen.
4. ______________ Refrigerator magnets hold reminder notes and display artwork and photographs.
5. ______________ Recyclers use magnets to sort iron from other types of metal.
6. ______________ Craftspeople use magnets to find dropped needles, pins, and other small parts.
7. ______________ Compasses use Earth's magnetism to find direction.
8. ______________ Electric can openers use a magnet to keep the lid from dropping down into the food.
9. ______________ Maglev trains use super conducting magnets between the track and the train so it "floats" above the track.
10. ______________ A cow magnet is swallowed by cattle to prevent "hardware disease," caused when cattle eat things like nails, bailing wire, or staples when grazing.
11. ______________ Magnets are used in dentistry to realign teeth into their correct place or spacing.
12. ______________ Food manufacturers use magnets to remove metal filings that may have dropped in food during production.

Name ______________________________ Date ______________

Magnetic Fields

Magnetic fields are the force fields generated around a magnet by moving electrical charges. We can see magnetic fields by placing a piece of paper over a magnet and lightly sprinkling iron filings on the paper. The outline of the magnet will appear and the filings will align themselves with the fields. Magnetic fields are actually three-dimensional, surrounding a magnet on all sides. Use the phrases in the word box to label each diagram. Then draw how the magnetic fields surround each set of magnets.

magnetic field around single magnet	magnetic fields around repelling ends
magnetic fields around attracting ends	

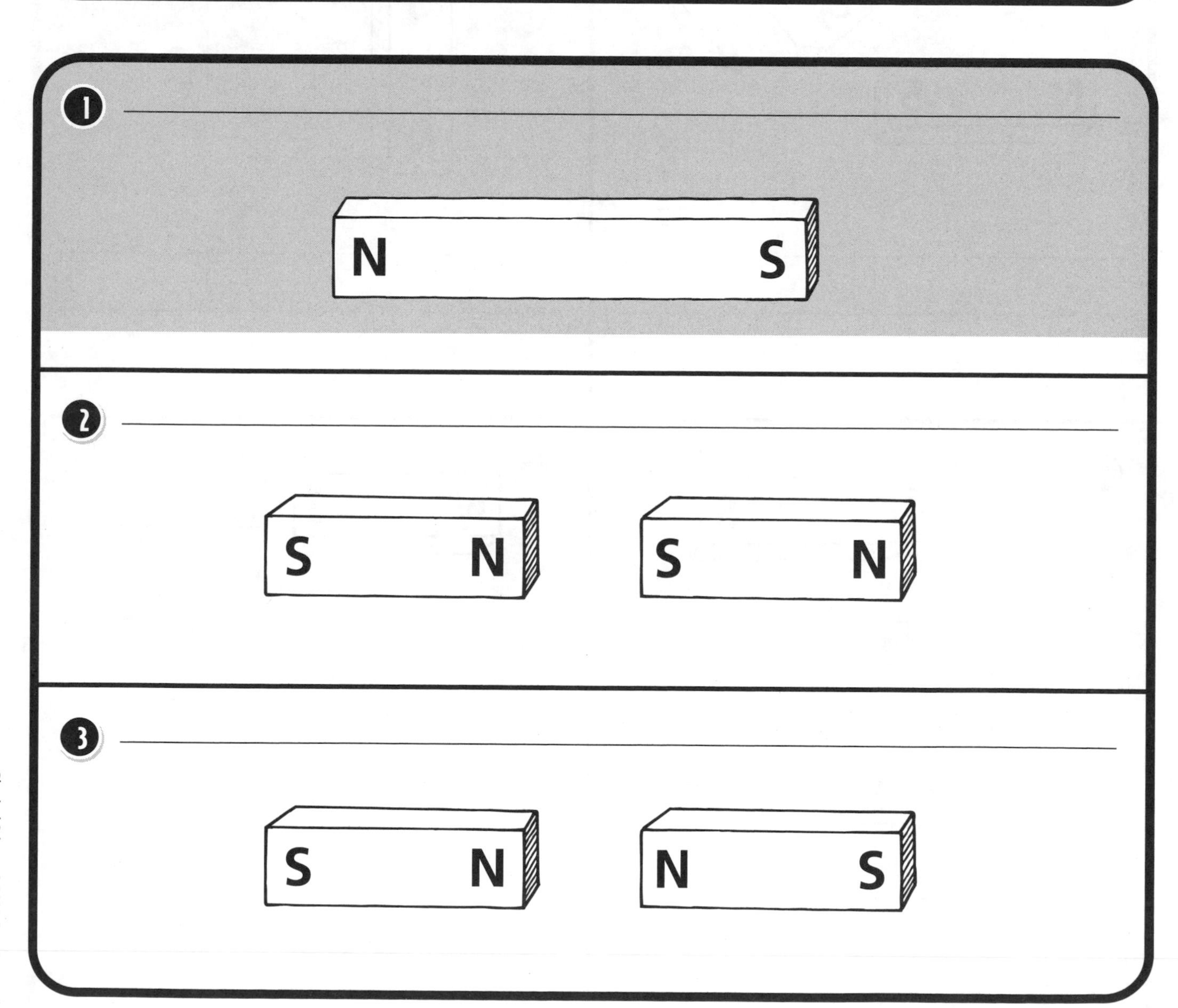

Name ______________________________ Date ______________

Attract or Repel?

Describe what will happen in each diagram.

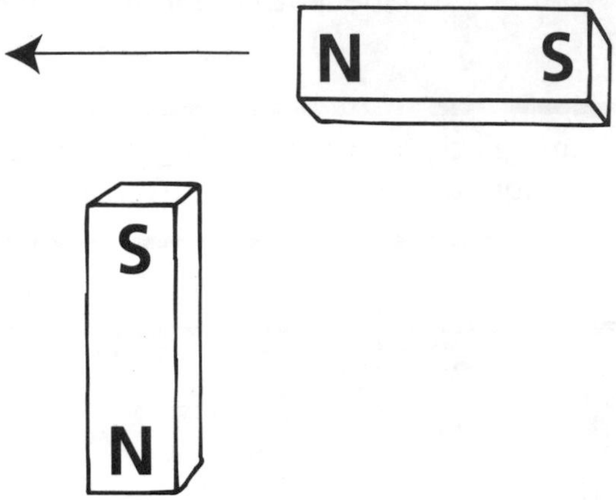

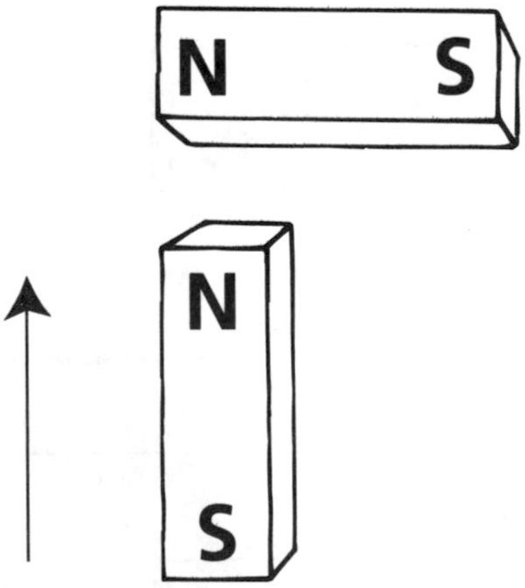

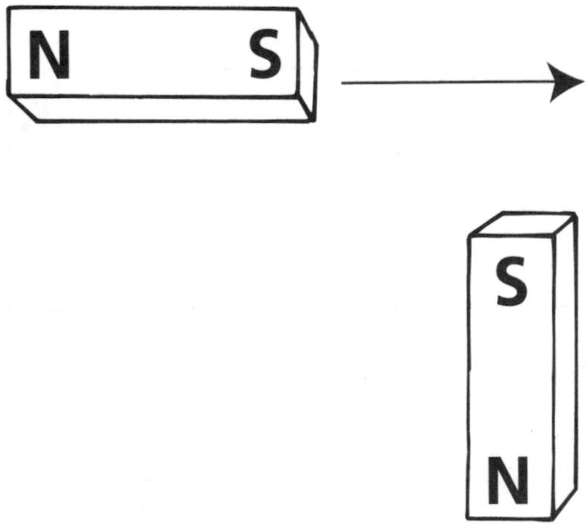

1. N S — S N

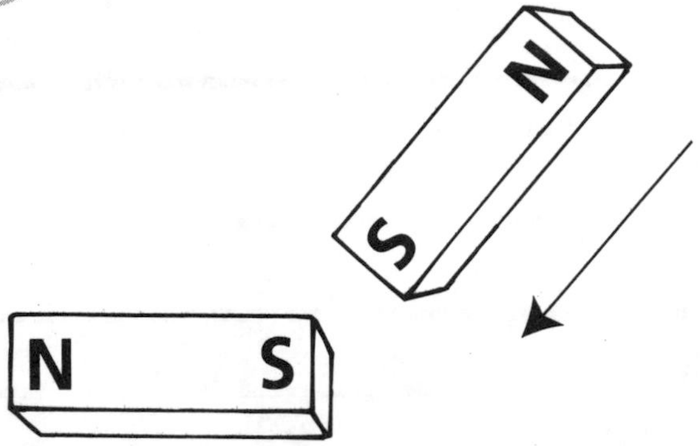

2. N S — S N

3. N S — N S

4. N S — S N

Name ______________________ Date ____________

Series Circuit

A **series circuit** is one in which the components are arranged end to end to form a single path for current. Use the following items and draw a series circuit in the box. Label the parts.

light bulb A	light bulb B	light bulb C
wire	switch	battery

Answer the questions about a series circuit.

1. If Bulb A burns out, will Bulb C still light? Explain. ______________________

   ____

2. What happens if the switch is open? ______________________

   What happens if it is closed? ______________________

3. What would happen if another battery is added to the circuit? ______________________

   ____

4. Is the wiring in your house a series circuit? How can you tell? ______________________

   ____

5. Give an example of something that has a series circuit, and tell how you know. ______________________

   ____

Name ______________________ Date ____________

Parallel Circuit

A **parallel circuit** is one in which two or more components are connected across two common points so there are separate conducting paths for the current. Use the following items and draw a series circuit in the box. Label the parts.

light bulb A	light bulb B	light bulb C
switch	battery	wire

Answer the questions about a series circuit.

1. If Bulb A burns out, will Bulb C still light? Explain. ______________________

____

2. What happens if the switch is open? ______________________

What happens if it is closed? ______________________

3. What would happen if another battery is added to the circuit? ______________________

____

4. Is the wiring in your house a parallel circuit? How can you tell? ______________________

____

5. Give an example of something that has a parallel circuit, and tell how you know. ______________________

____

Name ______________________________ Date ______________

Identifying Types of Circuits

Label each of the following diagrams as a **parallel** or **series** circuit. Use the terms in the word box to label the parts of each.

power	bulb	switch	wire

1. Type: ______________________
2. Type: ______________________

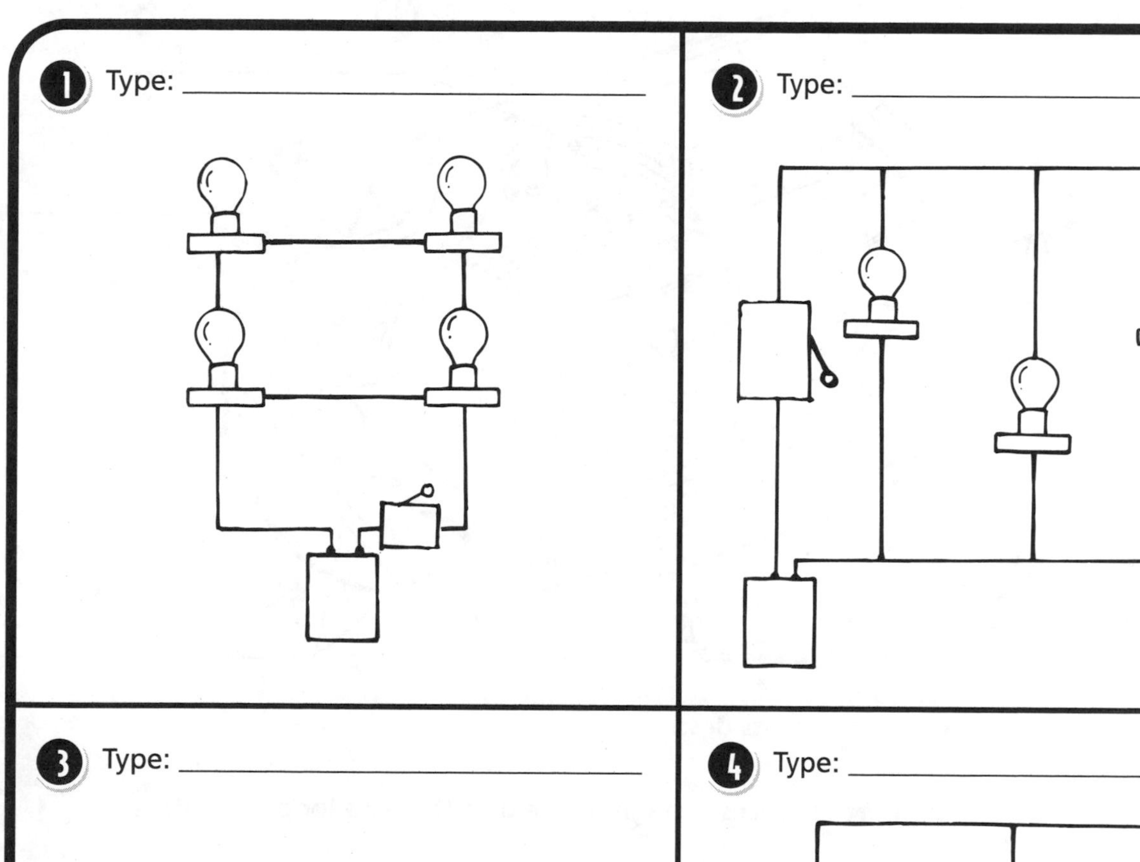

3. Type: ______________________

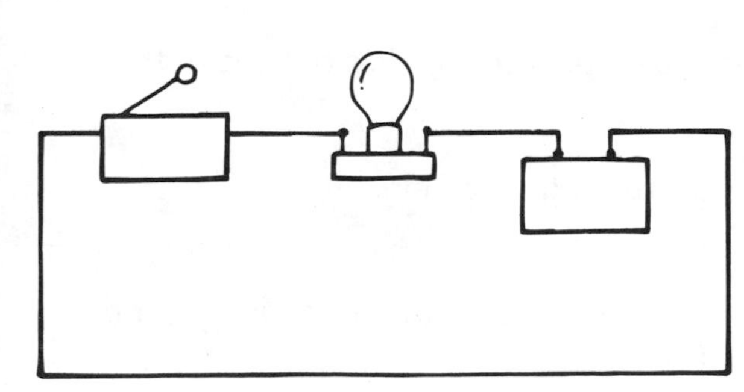

4. Type: ______________________

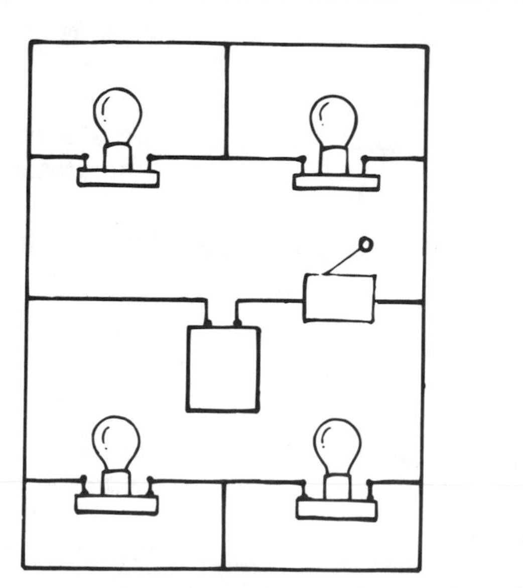

Name ______________________________ Date ______________

An Electric Motor

Electric motors are used in many appliances, tools, and toys. Use the terms in the word box to label the parts of an electric motor. Then use the same terms and match them to the descriptions below.

permanent magnet	armature	commutator	field coil
current source	brushes		

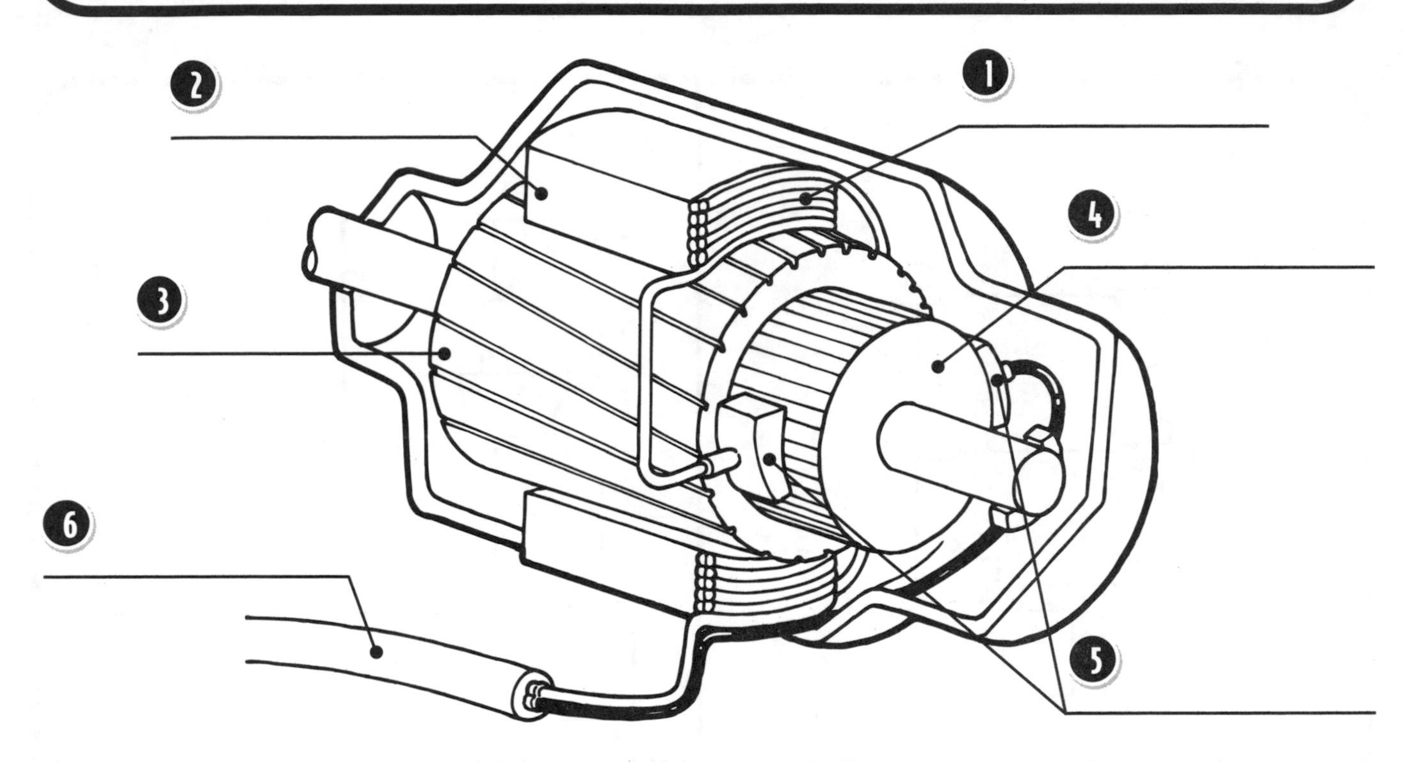

Match each term in the word box above to its description.

7. ______________ This part produces a magnetic field so the wire loop will rotate.

8. ______________ One is positive and one is negative. Each supplies current to the commutator.

9. ______________ This part strengthens the electromagnetic field.

10. ______________ This consists of many loops of wire wound around an iron cylinder. It rotates as current flows through the wire.

11. ______________ This part switches direction of current flow so the poles of the magnet are periodically reversed.

12. ______________ This provides the current to the coil so it in turn becomes a magnet.

Name ______________________ Date ______________

Comparing Magnetism and Electricity

Magnetism and electricity have many similarities, but each retains properties slightly different than the other. Use the phrases in the word box to complete the chart.

can be used to make magnetism	N attracts S	like poles repel
two opposite charges attract	+ attracts –	like charges repel
magnetic forces can attract and repel at a distance		two opposite poles attract
electric forces can attract and repel at a distance		can be used to make electricity

Magnetism	Electricity
N S	
____________________	____________________
____________________	____________________
____________________	____________________
____________________	____________________
____________________	____________________

Answer Key

The Scientific Method (page 5)

A. make a hypothesis; 2
B. observe and record; 4
C. interpret data; 5
D. test the hypothesis; 3
E. arrive at a conclusion; 6
F. identify the problem; 1

Scientific Definitions (page 6)

1. experiment
2. hypothesis
3. data
4. variable
5. theory
6. conclusion
7. control
8. procedure

Laboratory Safety (page 7)

1. secure loose clothing and hair
2. keep flammables away from flame
3. use scientific smelling when needed
4. handle glass and sharp materials carefully
5. keep workspace clean and organized
6. wear proper safety equipment

Equipment in the Laboratory (page 8)

1. funnel
2. test tube
3. tongs
4. beaker
5. balance
6. test tube clamp
7. ring stand
8. Erlenmeyer flask
9. Bunsen burner
10. graduated cylinder

Using Lab Equipment (page 9)

1. Erlenmeyer flask
2. beaker
3. test tube clamp
4. test tubes
5. graduated cylinder
6. funnel
7. tongs
8. balance
9. Bunsen burner
10. ring stand

Tools for Measurement (page 10)

1. balance — measures mass of an object or material
2. thermometer — measures temperature
3. spring scale — measures force
4. metric ruler — measures length
5. graduated cylinder — measures volume of liquids
6. calipers — measures thickness or small distances

Using a Balance Scale (page 11)

1. adjustment screw
2. riders
3. pointer
4. scale
5. beams
6. pan
7. 205.6 grams
8. 45.5 grams
9. 153.7 grams
10. 514.0 grams

Measuring Metric Length (page 12)

1. kilometer
2. milliimeter
3. centimeter
4. meter
5. kilometer
6. centimeter
7. meter
8. millimeter
9. meter
10. millimeter

Measuring Liquids (page 13)

1. 56 mL
2. 16.5 mL
3. 76 mL
4. 4.34 mL
5. 23.5 mL
6. 32 mL

Measuring Temperature (page 14)

1. 32 degrees
2. 0 degrees
3. 212 degrees
4. 100 degrees
5. 28.1°
6. -2.5°
7. 68°
8. 98.7°
9. 11.5°
10. 11°

Choosing Units of Measurement (page 15)

1. nm
2. mm
3. cm
4. m
5. km
6. µg
7. mg
8. g
9. kg
10. ms
11. mL
12. l
13. kg
14. µg
15. mg
16. g
17. km
18. l
19. nm
20. m
21. mL
22. cm
23. mm
24. ms

Scientific Notation (page 16)

1. 5.0×10^{6}
2. 1.2×10^{5}
3. 2.2×10^{7}
4. 6.0×10^{3}
5. 2.75×10^{8}
6. 4.5×10^{4}
7. 8.0×10^{-3}
8. 4.5×10^{-3}
9. 7.5×10^{-4}
10. 3.55×10^{-2}
11. 1.56×10^{-3}
12. 4.2×10^{-5}

The Atom (page 17)

1. atom
2. electron
3. electron orbit
4. neutron
5. nucleus
6. proton
7. electron
8. nucleus
9. neutron
10. electron orbit
11. atom
12. proton

Elements (page 18)

Gold
shiny
conducts heat and electricity
can be hammered into sheets
can be pulled into wires

Helium
nonconducting
occurs naturally as a gas
lighter than air

Mercury
nonconducting
occurs naturally as a liquid
poisonous

Copper
shiny
conducts heat and electricity
can be hammered into sheets
can be pulled into wires

Silver
shiny
conducts heat and electricity
can be hammered into sheets
can be pulled into wires

Oxygen
nonconducting
required for combustion
occurs naturally as a gas

Setting the Table (page 19)

1. transition
2. noble
3. metals
4. alkali metals
5. periods
6. atomic number
7. families
8. alkaline earth metals
9. rare earth metals
10. naturally

Decoding the Elements (page 21)

1. atomic number
2. element symbol
3. element name
4. atomic mass
5. element symbol
6. element name
7. atomic number
8. atomic mass

Elements and Their Symbols (page 22)

1. copper
2. calcium
3. lead
4. iodine
5. potassium
6. carbon
7. tin
8. nickel
9. aluminum
10. gold
11. S
12. F
13. Xe
14. W
15. Cr
16. Si
17. Hg
18. Na
19. Pt
20. As

Real-World Applications (page 23)

1. tungsten
2. platinum
3. sulfur
4. nickel
5. neon
6. chromium
7. chlorine
8. helium
9. lead
10. tin

Atoms by Element (page 24)

[See diagram on page 123.]

Atomic Ions (page 25)

1. Ion: cation Charge: positive
2. Ion: anion Charge: negative
3. Ion: cation Charge: positive
4. Ion: anion Charge: negative
5. Ion: anion Charge: negative
6. Ion: anion Charge: negative

Symbolic Atoms (page 26)

1.	12	24	0	12	12	12
2.	9	19	-1	9	10	10
3.	16	32	-2	16	16	18
4.	1	1	0	1	0	1
5.	4	9	+2	4	5	2
6.	11	23	+1	11	12	10
7.	19	39	0	19	20	19

Elemental Definitions Part I (page 27)

1. carbon	2. selenium	3. arsenic
4. phosphorus	5. flourine	6. oxygen
7. helium	8. bismuth	9. bromine
10. chlorine		

Elemental Definitions Part II (page 28)

1. nitrogen	2. sulfur	3. hydrogen
4. cobalt	5. calcium	6. neon
7. boron	8. lithium	9. polonium
10. zirconium		

Properties of Metals and Nonmetals (page 29)

Properties of Metal Elements	Examples
malleable	zinc
lustrous	nickel
ductile	gold
conductor	titanium
metallic bonding	Thallium
gives away electrons in chemical reactions	

Properties of Nonmetal Elements	Examples
covalent bonding	selenium
forms negative ions	boron
brittle	argon
nonconductor	helium
gaseous at room temperature	phosphorus
receives electrons in chemical reactions	

Molecules (page 30)

1. an ammonia molecule	2. a water molecule
3. hydrogen atoms	4. nitrogen atom
5. oxygen atom	6. hydrogen atoms
7. a carbon dioxide molecule	8. oxygen atom
9. carbon atom	

Three States of Matter (page 31)

Solid
takes up space
has mass
has a shape of its own
strong bonds between molecules
hard to deform
does not expand
molecule movement is smallest
has definite volume

Liquid
takes up space
has mass
takes shape of container
has definite volume
weak bonds between molecules
spreads in direction of gravity
does not expand

Gas
takes up space
has mass
takes shape of container
has no definite volume
virtually no bonds between molecules
spreads in all directions
expands
molecule movement is greatest

Physical and Chemical Properties (page 32)

Physical Property	Chemical Property
color	flammability
density	supports combustion
solubility	neutralizes a base
taste	neutralizes an acid
melting point	reacts with an acid
hardness	reacts to oxygen
boiling point	reacts with water to form gas
luster	reacts with base to form water
odor	electromotive

Physical Changes (page 33)

1. condensation	2. sublimation
3. solid	4. evaporation
5. density	6. fusion
7. suspension	8. crystallization
9. solution	10. substances
11. liquid	12. physical
13. gas	14. plasma

Chemical Changes (page 34)

1. endothermic	2. catalysts	3. reactants
4. precipitate	5. enzyme	6. corrosion
7. combustion	8. exothermic	9. reaction
10. products	11. acid	12. flammable
13. base		

Identifying Physical and Chemical Changes (page 35)

1. chemical	2. physical	3. chemical
4. physical	5. physical	6. physical
7. chemical	8. physical	9. chemical
10. chemical	11. physical	12. chemical

Mixtures, Compounds, and Solutions (page 36)

Mixtures	Compounds	Solutions
air	water	rubbing alcohol
blood	carbon dioxide	sugar water
sand	ammonia	salt water
	sugar	carbonated soft drink
	alcohol	humidity
	table salt	brass
		milk
		coffee
		ink

Separation of Mixtures (page 37)

1. evaporation	2. sifting or filtering
3. weight	4. magnetism
5. evaporation or weight	6. weight
7. magnetism	8. sifting or filtering
9. evaporation	10. weight

Homogeneous and Heterogeneous Mixtures (page 38)

Heterogeneous Mixtures	Homogeneous Mixtures
spaghetti sauce	flat soda pop
city air	sugar
soil	aluminum foil
beach sand	sugar water
vegetable soup	paint
oil and vinegar salad dressing	mayonnaise
chocolate chip ice cream	alcohol
	black coffee

Solutions, Colloids, and Suspensions (page 39)

Suspension

Properties:
large particles
murky or opaque
separated by filtration
scatters light

Examples:
muddy water
oil and vinegar dressing
pulpy orange juice

Colloid

Properties:
medium particles
does not settle out
scatters light
particles remain evenly distributed
murky or opaque

Examples:
whipped cream
milk
fog

Solution

Properties:
minute particles
does not settle out
transparent
solute is dissolved into a solvent
paticles remain evenly distributed

Examples:
salt water
sugar water
rubbing alcohol

Types of Chemical Bonds (page 40)

1. valence electrons	2. covalent bond
3. ionic bond	4. anion
5. cation	6. properties
7. metallic bonding	8. single covalent bond
9. double covalent bond	10. hydrogen bond

Identifying Chemical Bonds (page 41)

1. sodium	chlorine	ionic
2. carbon	oxygen	covalent
3. hydrogen	oxygen	covalent
4. potassium	oxygen	ionic
5. sodium	fluorine	ionic
6. carbon	hydrogen	covalent
7. sulfur	oxygen	covalent
8. magnesium	oxygen	ionic
9. hydrogen	chlorine	covalent
10. nitrogen	oxygen	covalent
11. iron	chlorine	ionic
12. phosphorus	oxygen	covalent

Writing in Code (page 42)

1. 10 carbon atoms, 14 hydrogen atoms, 2 nitrogen atoms
2. 3 carbon atoms, 9 hydrogen atoms, 1 nitrogen atom
3. 14 carbon atoms, 8 hydrogen atoms, 5 oxygen atoms, 2 nitrogen atoms
4. 10 carbon atoms, 12 hydrogen atoms, 1 oxygen atom, 2 nitrogen atoms

Number of Atoms by Formula (page 43)

1. 2; sodium chloride	2. 4; hydrogen peroxide
3. 4; mercurous chloride	4. 5; iron oxide
5. 8; phosphoric acid	6. 6; potassium carbonate
7. 3; calcium chloride	8. 6; ammonium bromide
9. 6; copper sulfate	10. 7; sulfuric acid
11. 6; sodium sulfite	12. 5; silver nitrate

Chemical Formulas by Name (page 44)

1. CO	2. SO_3	3. NaCl
4. CCl_4	5. N_2O	6. N_2O_5
7. CO_2	8. PCl_3	9. Ma_3N_2
10. AlI_3	11. $MgBr_2$	12. SO_3
13. PCl_5	14. AlCl	15. H_2O

Forces and Motion (page 45)

1. friction	2. mass	3. centripetal
4. gravity	5. speed	6. acceleration
7. velocity	8. kinetic	9. inertia
10. newton	11. weight	12. momentum

Balanced and Unbalanced Forces (page 46)

1. balanced	2. gravity
3. table	4. rest
5. unbalanced	6. gravity
7. accelerate	8. balanced
9. inertia	10. zero
11. state of motion	12. equilibrium

First Law of Motion (page 47)

1. The force of smacking the bottle is greater than the force holding the ketchup inside.

The ketchup keeps moving.

2. The force of the bodies traveling forward is greater than the force holding them in their seats.

The bodies keep moving.

3. The force of the rock against the skateboard is greater than the force of the skateboard.

The skateboarder keeps moving.

4. The force of tripping on the wrinkle is greater than the forward movement of his feet.

Jack and water keep moving.

Second Law of Motion (page 48)

1. Amy uses less mass so using the same amount of force, she will accelerate faster than her dad.

2. If they run at same speed, Tony (more mass) will hit the dummy with more force and cause it to accelerate faster.

3. Smaller car will accelerate faster than truck because smaller mass accelerates faster and requires less force to move the same acceleration.

Third Law of Motion (page 49)

1. Action: Girl pushes boy..
Reaction: Boy pushes back on her hands with equal force causing both kids to roll away.

2. Action: Golf club hits ball.
Reaction: Ball hits back on the club (that's why clubs have cushions and shock absorbers to absorb the reaction force of impact).

3. Action: Frog legs push down on lily pad.
Reaction: Lily pad pushes back on frog's legs allowing it to go forward.

Identifying Newton's Laws (page 50)

1. First Law of Motion: Rocket in motion will stay in motion unless gravity becomes greater than thrust.
2. Third Law of Motion: Thrust of engines push down on Earth, Earth pushes back which allows shuttle to go upward.
3. Second Law of Motion: More massive player will hit with greater force.
4. Third Law of Motion: Action- big player hits smaller player. Reaction- smaller player hits big player with equal force in opposite direction.
5. First Law of Motion: A boat in motion will stay in motion unless friction/gravity become greater.
6. Third Law of Motion: Action force- paddles push on water. Reaction force- water pushes back on paddles allowing boat to move.

Determining Speed and Velocity (page 51)

1. 30 kilometers per hour	2. 6.67 kilometers per hour
3. 600 miles per hour	4. 65.5 miles per hour
5. 90.5 kilometers per hour	6. 120 miles per hour
7. 1200 kilometers per hour	8. 2.5 miles per hour

Collisions (page 52)

1. [see diagrams on page 123]
Both are moving and both have same mass.
They move away at an angle because they collide at an angle.

2. [see diagrams on page 123]
The marble has more mass than the Ping Pong ball.
They move away at an angle because they collide at angle.

3. [see diagrams on page 123]
Ball #1 stops because it transferred its momentum to ball #2.
Ball #1 transfers some of its momentum to ball #2.

Gravity (page 53)

1. mass
2. weight
3. gravity
4. inertia
5. Newton
6. acceleration
7. tides
8. Einstein
9. friction
10. altitude

Effects of Gravity (page 54)

1. Use gravity on the downward swing. Overcome gravity when he swings his body upward.
2. Use gravity to pull the ball down into the basket. Overcome gravity by aiming and pushing the shot to go into the basket.
3. Use gravity to pull her down toward the water. Overcome gravity by springing on a dive board.
4. Use gravity to pull the body toward Earth. Overcome gravity by slowing descent with glider.

Friction (page 55)

<u>Friction Is Used</u>
rubbing two sticks together starts a fire
build a house of cards
press on a car's brakes
walk across the road
grate cheese
smooth wood with sandpaper
walk across a wood floor in shoes
pedal your bicycle
rub your hands together to warm them

<u>Friction Is Reduced</u>
a door hinge is oiled
lotion helps remove a tight gold ring
a dolphin glides through the water
grease a bicycle chain
skate across the ice
butter a cake pan
walk across a wood floor in socks
slide down a snowy hill
swan dive into a pool
a canoe glides down a river

Air Resistance (page 56)

1. Reducing resistance.
How? The streamlined body of the car allows it to move through the air with less resistance.
2. Using resistance.
How? The surface of the glider increases air resistance and slows down the effect of gravity.
3. Using resistance.
How? The shape of the kite increases air resistance and uses it to keep the kite floating on air currents.
4. Reducing resistance.
How? The tucked position of the skier and the shape of the skis and uniform reduce air resistance so the skier goes faster and farther.

Potential vs. Kinetic Energy (page 57)

1. potential energy
2. potential energy
3. potential energy
4. kinetic energy
5. kinetic energy
6. kinetic energy
7. potential energy
8. kinetic energy
9. kinetic energy
10. potential energy
11. kinetic energy
12. potential energy

Calculating Work (page 58)

1. 6 joules
2. 13 joules
3. 2400 joules
4. 67.5 joules
5. 8.0 joules
6. 16.0 joules
7. 1.5 meters
8. 20 newtons

Pendulum Swings (page 59)

1. fixed point
2. bob
3. period
4. gravity
5. simple pendulum
6. Foucault
7. Galileo
8. amplitude
9. frequency
10. North Pole
11. equator
12. metronome

The Period of a Pendulum (page 60)

[see graph on page 124]

Simple Machines (page 61)

1. lever or wedge
2. inclined plane
3. wedge
4. wheel and axle
5. screw
6. pulley
7. lever
8. lever
9. wheel and axle

Functions of Simple Machines (page 62)

1. work
2. screw
3. pulley
4. gear
5. distance
6. wheel and axle
7. force
8. lever
9. wedge
10. inclined plane

Identifying Parts as Simple Machines (page 63)

1. lever, wedge
2. lever, wedge
3. Screw, wheel and axle
4. inclined plane, wheel and axle
5. wedge
6. pulley

Levers at Work (page 64)

1. distance you use
2. force
3. load
4. distance lever uses
5. fulcrum
6. distance you use
7. distance lever uses
8. load
9. fulcrum
10. force

Three Classes of Levers (page 65)

Type of Lever: third class

1. load
2. effort
3. fulcrum

Type of Lever: first class

4. load
5. fulcrum
6. effort

Type of Lever: second class

7. fulcrum
8. load
9. effort

Classes of Levers (page 66)

1. second class
2. second class
3. third class
4. first class
5. first or third class
6. second class
7. second class
8. third class
9. third class

Inclined Planes (page 67)

1. The ladder helps us climb up a height at an easier angle. With the ladder at an angle, I can't climb as high, but the climb is easier.
2. The stairs help me climb to a higher place. The angle of the stairs is easier to climb than a steep ladder.
3. These stairs are not steep. I don't have to work so high to climb up, but I don't climb as high either.
4. The ramp lets me roll something into the car. It is easier to roll the object up the slope than it is to lift the object straight up into the car.

Mechanical Advantage of Inclined Planes and Levers (page 68)

1. MA = 4
2. MA = 7
3. MA = 5
4. MA = 3

Pulleys (page 69)

1. pulley
2. distance to move load
3. length of rope pulled
4. force
5. distance to move load
6. pulley
7. length of rope pulled
8. force

Wheels and Axles (page 70)

1. force
2. distance axle uses
3. distance you use
4. small wheel is difficult to turn
5. large wheel makes turning easier
6. wheel and axle
7. wheel and axle
8. reduce friction
9. wheel and axle
10. reduce friction
11. wheel and axle

Mechanical Advantage of Pulleys, Wheels, and Axles (page 71)

1. ma = 4
2. ma = 2
3. ma = 5
4. ma = 40

Wedges (page 72)

1. attach
2. tighten
3. separate
4. separate
5. lift
6. separate
7. attach
8. separate
9. tighten
10. tighten
11. separate or lift
12. separate

Screws (page 73)

1. drill
2. inclined plane
3. turned
4. wheel and axle
5. wedge
6. force
7. threads
8. distance
9. jackscrew
10. handle
11. axle

Classifying Simple Machines (page 74)

Levers
hammer
bottle opener
seesaw
fishing pole
scissors

Screws
nut and bolt
hand drill
light socket

Inclined Planes
step stool
staircase
ladder
ramp

Wedges
scissors
zipper
front teeth
knife

Pulleys
flagpole
crane
window blinds

Wheel and Axles
bicycle gears
door knob
faucet handle
wrench and pipe

Uses for Simple Machines (page 75)

1. pulley
2. screw
3. lever
4. wedge
5. inclined plane
6. pulley
7. screw
8. lever
9. inclined plane
10. wheel and axle
11. wedge
12. wheel and axle or pulley

The Various Forms of Energy (page 76)

1. potential energy
2. electrical energy
3. nuclear energy
4. kinetic energy
5. thermal energy
6. radiant energy
7. chemical energy
8. forms
9. sound energy
10. mechanical energy

Types of Energy (page 77)

1. mechanical
2. radiant
3. chemical
4. nuclear
5. sound
6. thermal
7. electrical

Energy of One Kind or Another (page 78)

1. electrical energy
2. mechanical energy
3. most potential energy
4. potential energy changing to kinetic energy
5. most kinetic energy
6. thermal energy

From One Form to Another (page 79)

1. radiant to chemical
2. radiant to thermal
3. electrical to radiant
4. mechanical to electrical
5. chemical to thermal
6. electrical to sound
7. thermal to mechanical
8. electrical to mechanical

Energy Transfers (page 80)

1. sound
2. thermal
3. electrical
4. nuclear
5. chemical
6. mechanical
7. radiant
8. chemical
9. thermal
10. mechanical
11. electrical

Classifying Potential or Kinetic Energy (page 81)

Potential Energy
standing at the top of a slide
wind up for the pitch
a battery
juice in an orange
an unburned lump of coal
frog sitting on a lily pad
book on a high shelf
a parked car

Kinetic Energy
move down a slide
throw a curve ball
execute a swan dive
move downhill in a roller coaster
frog leaping into the water
roll down a grassy hill
book falls from a high shelf
a speeding car

An Electric Generator (page 82)

1. drive shaft
2. wheel and axle
3. magnet
4. wire coil
5. source of mechanical energy
6. wheel and axle
7. drive shaft
8. wire coil
9. magnet
10. source of mechanical energy

Energy from a Nuclear Reactor (page 83)

1. shield
2. transformer
3. coolant
4. control rods
5. reaction chamber
6. turbine generator
7. heat exchanger
8. fuel
9. moderator

Parts of a Nuclear Reactor (page 84)

1. turbine generator
2. transformer
3. heat exchanger
4. coolant
5. control rods
6. moderator
7. fuel
8. shield
9. reaction chamber

Alternative Energy Sources (page 85)

1. solar energy
2. windmill
3. wind
4. geothermal
5. hotspots
6. wind technology
7. sun
8. tidal
9. hydroelectric
10. cells
11. efficient

Terms of Light (page 86)

1. refraction
2. reflection
3. photon
4. crest
5. light
6. wavelength
7. visible light spectrum
8. trough
9. prism
10. wave velocity
11. hertz
12. frequency

Diagram of a Wave (page 87)

1. equilibrium
2. amplitude
3. wavelength
4. trough
5. crest
6. wavelength
7. trough
8. crest
9. amplitude
10. equilibrium

Light Waves (page 88)

1. gamma rays
2. X-rays
3. ultraviolet rays
4. visible light
5. infrared rays
6. microwaves
7. short radio waves
8. long radio waves

Uses of Electromagnetic Energy (page 89)

Gamma Rays
kills organisms that spoil food
treats some cancers

X-Rays
shows cavities in teeth
shows breaks in bones

Ultraviolet Rays
creates a tan
used to kill germs
causes a sunburn

Visible Light
allows us to see
creates a rainbow

Infrared Rays
shows heat loss in buildings
fire's heat
sun's heat

Microwaves
portions of phone calls
used to cook food
radar

Radio Waves
television signals
radio signals
cell phone signals
maritime communication

White Light Spectrum (page 90)

1. red; 780–622 nm
2. orange; 622–597 nm
3. yellow; 597–577 nm
4. green; 577–492 nm
5. blue; 492–475 nm
6. indigo; 475–445 nm
7. violet; 445–390 nm

Light Rays and Concave Lenses (page 91)

[see diagrams on page 124]

1. object	2. image
3. focal point	4. lens
5. object	6. image
7. lens	8. focal point

Light Rays and Convex Lenses (page 92)

[see diagrams on page 125]

1. object	2. lens
3. focal point	4. inverted image
5. object	6. lens
7. focal point	8. inverted image

Reflection (page 93)

[see diagrams on page 125]

Refraction (page 94)

1. light ray	2. incident angle
3. normal	4. surface
5. refractive angle	6. light ray
7. incident angle	8. normal
9. surface	10. refractive angle

Classify Refraction Materials (page 95)

Opaque	Translucent
light does not pass through	light scatters
aluminum foil	textured glass
construction paper	wax paper
mud	oiled paper
velvet	silty water
cardboard	sheer fabric
plywood	tracing paper
	prism

Transparent

light passes through
plastic wrap
window glass
clear water
car windshield

Diagram of a Sound Wave (page 96)

1. sound source	2. compression
3. rarefaction	4. wavelength
5. equilibrium	6. equilibrium
7. rarefaction	8. sound source
9. compression	10. wavelength

How Do These Sound? (page 97)

1. decibel	2. acoustics
3. loudness	4. octave
5. vacuum	6. ultrasonic
7. amplitude	8. Doppler
9. pitch	10. compression
11. rarefaction	12. harmony

Sound on a Frequency Scale (page 98)

1. speed of sound	2. longitudinal
3. amplitude	4. Doppler effect
5. pitch	6. vibrates
7. decibel	8. ultrasonic
9. compression	10. rarefaction
11. fluctuations	12. intensity

Wave Velocity (page 99)

1. 8.5 2. 420 3. 769.5 4. 924

How Loud Is Too Loud? (page 100)

Answers will vary. Possible answers include:

1. 140 decibels; hearing damage
2. 70 decibels
3. 40 decibels
4. 10 decibels
5. 20 decibels
6. 60 decibels
7. 96 decibels; hearing damage
8. 110 decibels; hearing damage
9. 80 decibels; hearing damage
10. 94 decibels; hearing damage
11. 80 decibels; hearing damage
12. 70 decibels
13. 80 decibels; hearing damage
14. 70 decibels

Electrical Terms (page 101)

1. lightning
2. volt
3. resistance
4. alternating
5. electric current
6. conductivity
7. circuit
8. watt
9. ammeter
10. direct
11. static electricity
12. kilowatt

Magnetic Terms (page 102)

1. lodestone
2. repel
3. compass
4. electromagnet
5. magnet
6. attract
7. magnetism
8. pole
9. magnetic field
10. Gauss meter

Types of Magnets (page 103)

1. horseshoe magnet
2. disk magnet
3. bar magnet
4. cylindrical magnet
5. U-shaped magnet
6. bar magnet
7. cow magnet
8. ring magnet
9. electromagnet

How Magnets Are Used (page 104)

1. used for detection
2. used as separators
3. use magnetic repulsion
4. hold or attract objects
5. used as separators
6. hold or attract objects
7. used for detection
8. hold or attract objects
9. use magnetic repulsion
10. hold or attract objects
11. hold or attract objects
12. used as separators

Magnetic Fields (page 105)

[see diagrams on page 126]

Attract or Repel? (page 106)

1. As the moving magnet approaches the end of the still magnet, the still one will swivel away because the two ends repel each other.
2. As the moving magnet gets close to the other magnet, it will attract and pull up the other magnet because the two ends attract each other.
3. As the moving magnet approaches the end of the still magnet, the still magnet will swivel up and away because the ends repel.
4. As the moving magnet nears the still magnet, the still magnet will back away because the two ends repel.

Series Circuit (page 107)

[see diagrams on page 127]

Parallel Circuit (page 108)

[see diagrams on page 128]

Identifying Types of Circuits (page 109)

1. series
2. parallel
3. series
4. parallel

Students should label examples for each term in each diagram.

An Electric Motor (page 110)

1. field coil
2. permanent magnet
3. armature
4. commutator
5. brushes
6. current source
7. permanent magnet
8. brushes
9. field coil
10. armature
11. commutator
12. current source

Comparing Magnetism and Electricity (page 111)

<u>Magnetism</u>

two opposite poles attract
N attracts S
like poles repel
magnetic forces can attract and repel at a distance
can be used to make electricity

<u>Electricity</u>

two opposite charges attract
+ attracts -
like charges repel
electric forces can attract and repel at a distance
can be used to make magnetism

Atoms by Element (page 24)

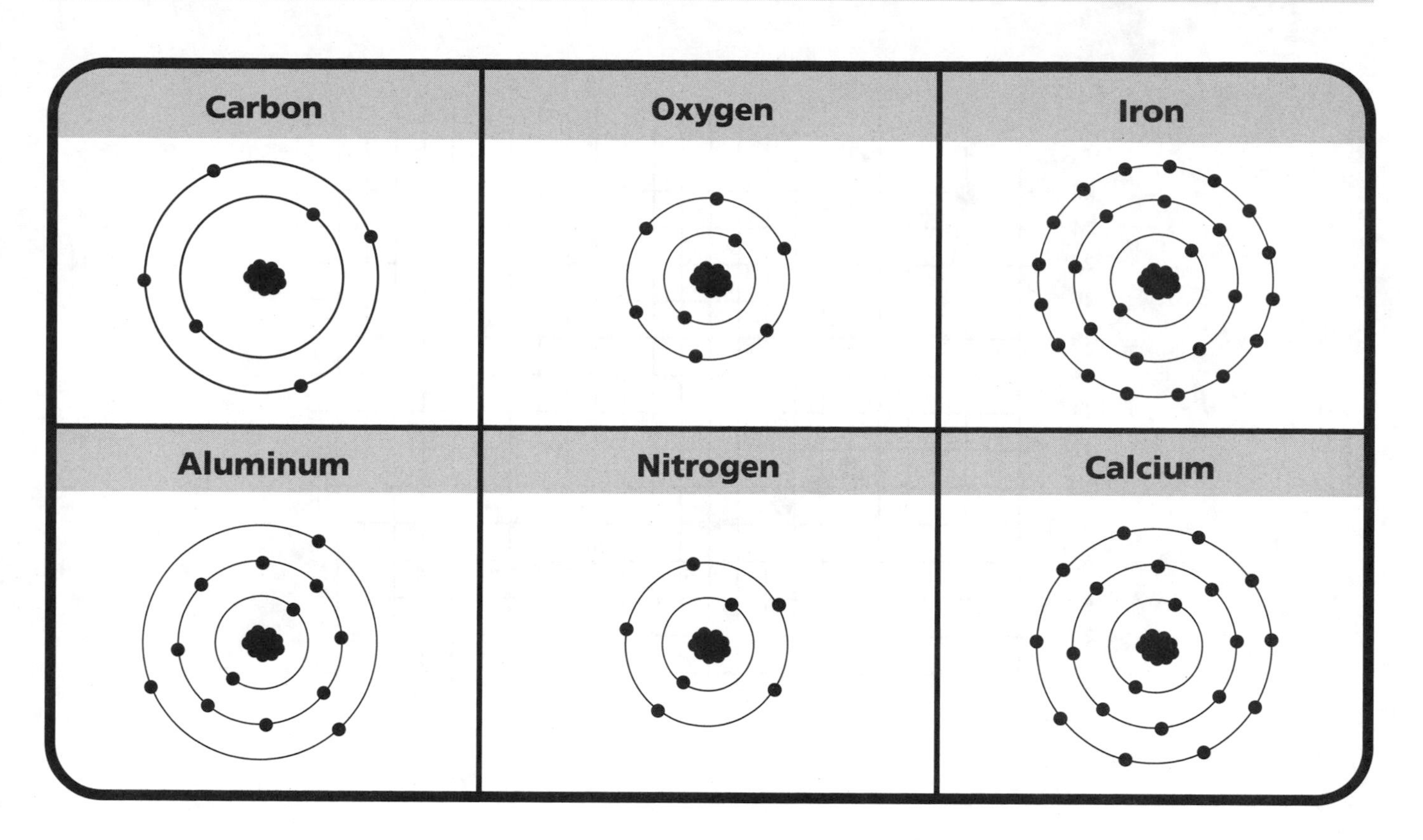

Collisions (page 52)

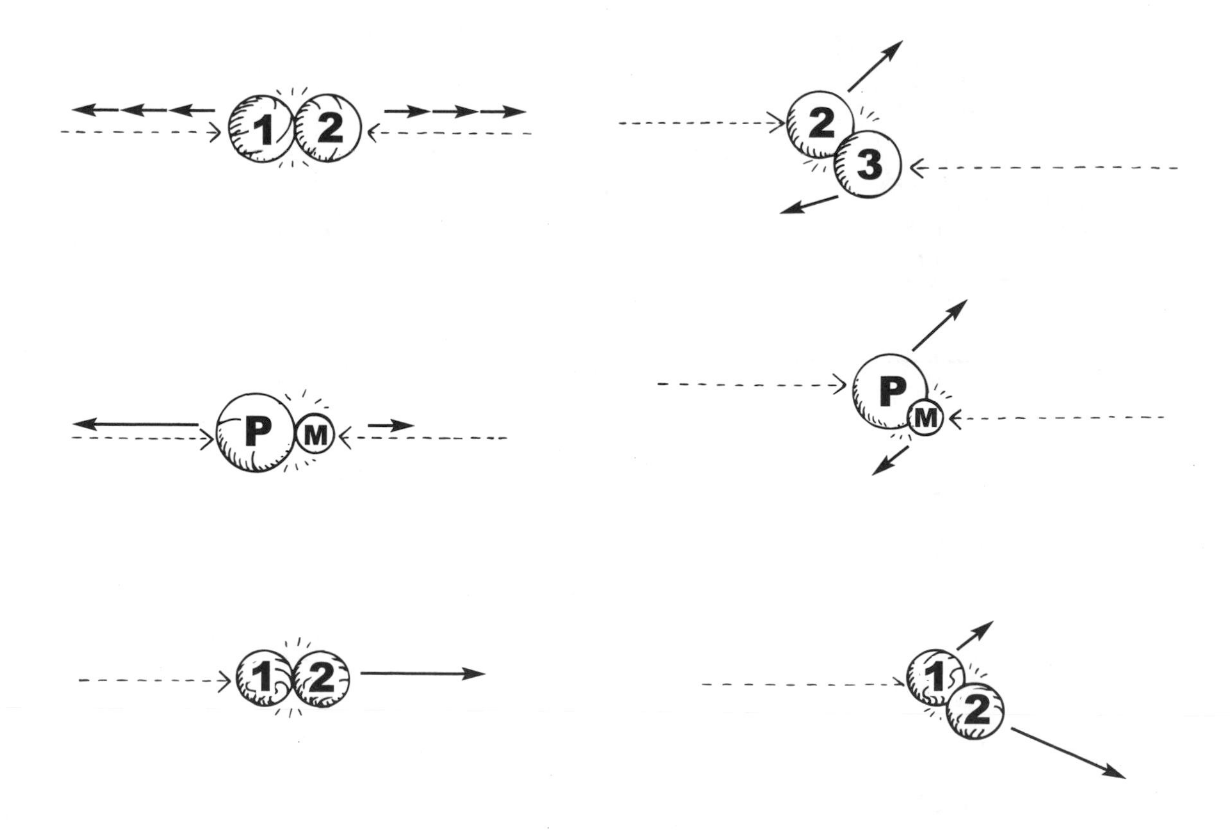

The Period of a Pendulum (page 60)

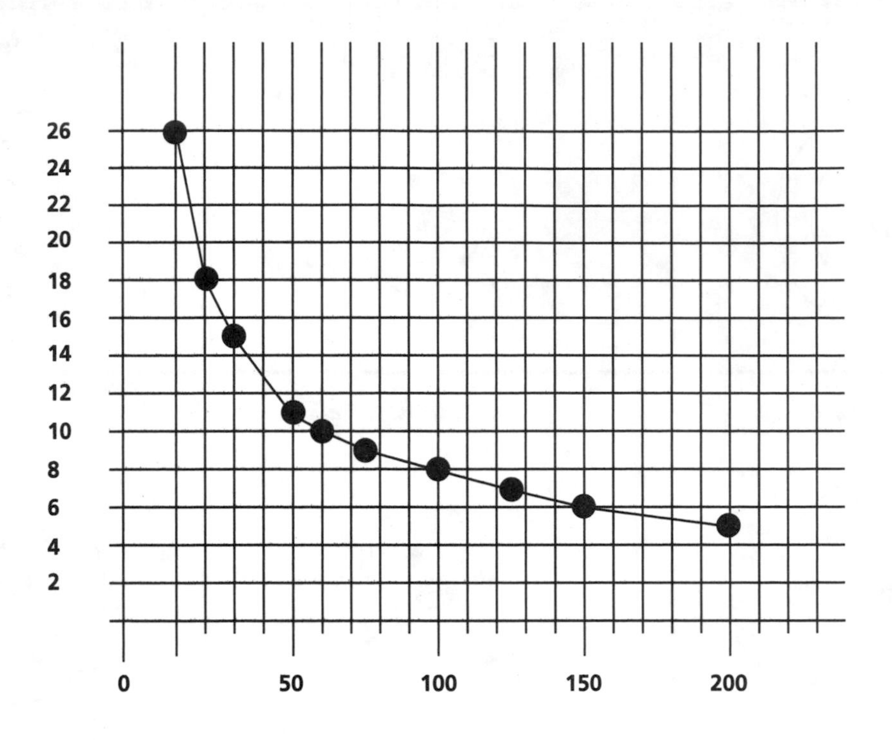

Light Rays and Concave Lenses (page 91)

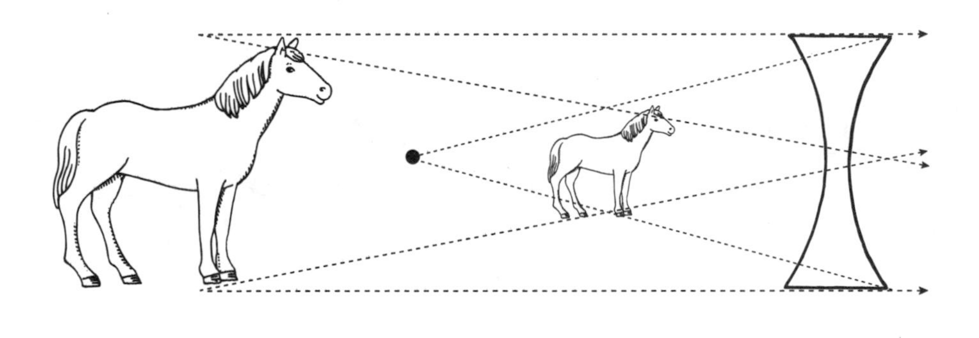

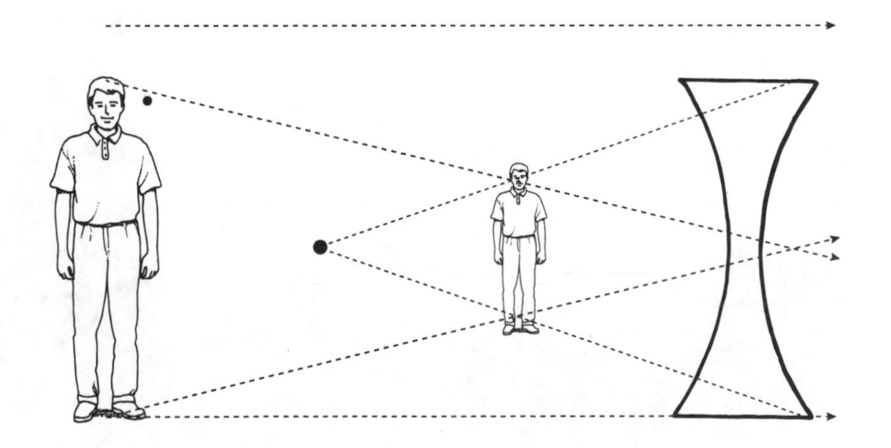

Light Rays and Convex Lenses (page 92)

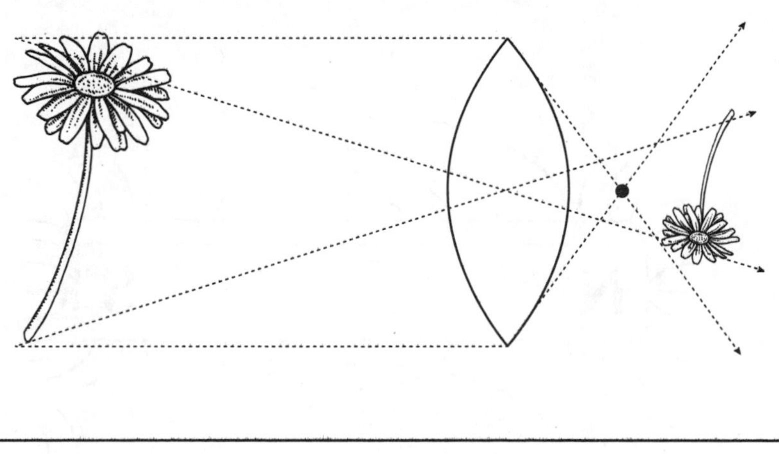

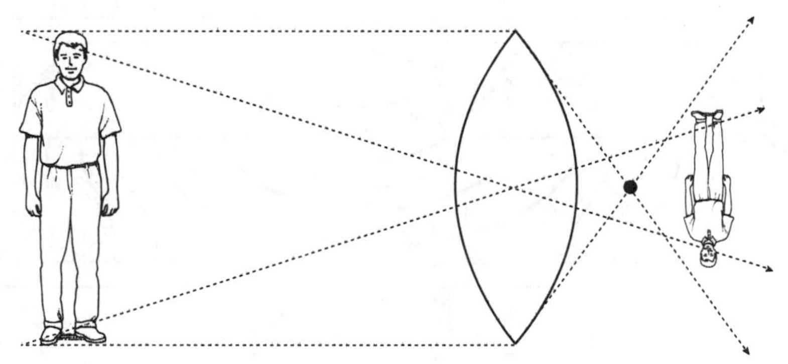

Reflection (page 93)

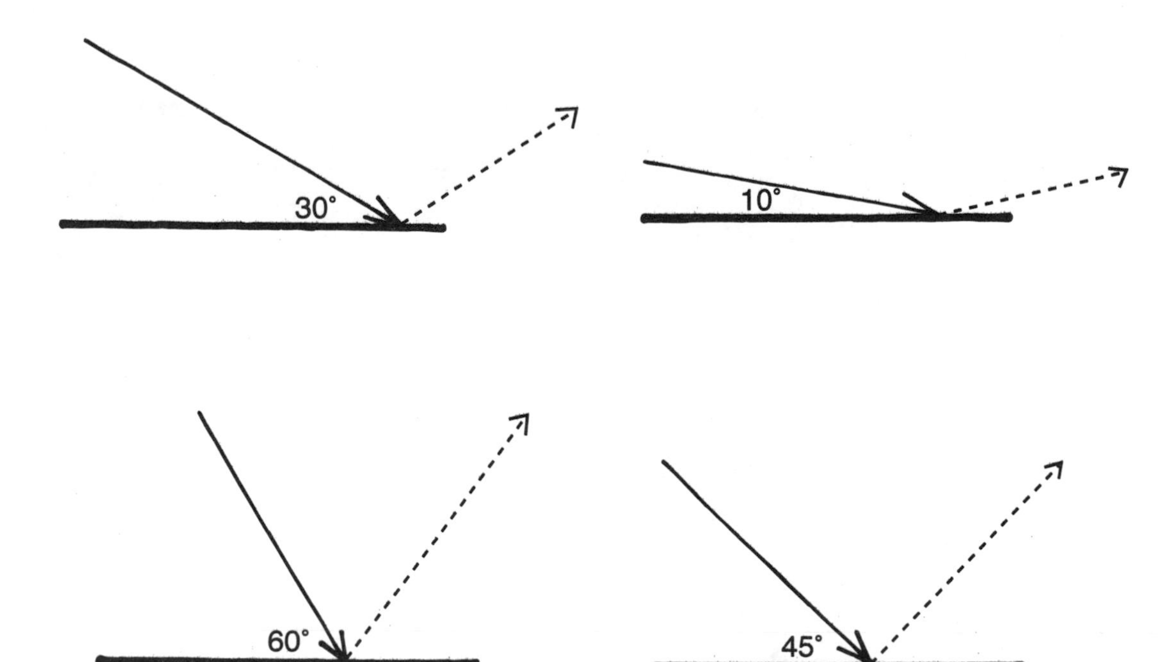

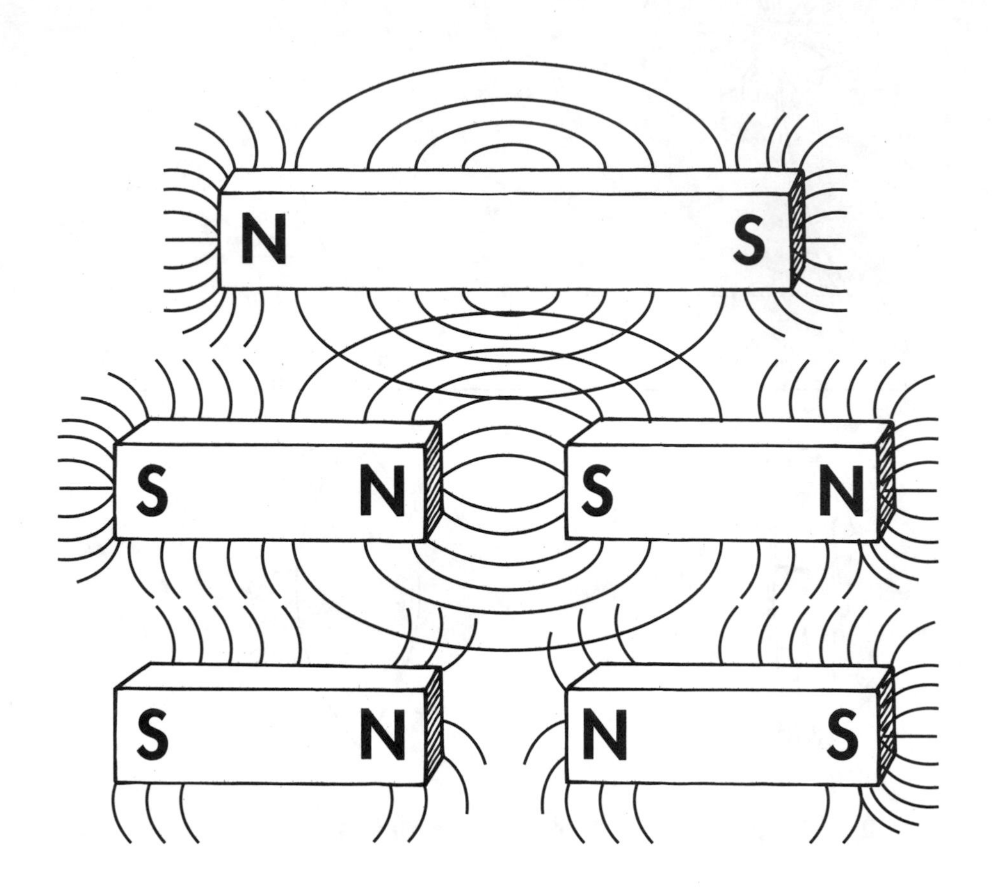
N
S
S
N
S
N
S
N
N
S

Series Circuit (page 107)

1. No, because there is a break in the circuit.
2. The lights won't light because the circuit is broken. The lights will light because now it is closed.
3. The lights would be brighter.
4. No, because if one light goes out, they don't all go out.
5. Possible answer: Christmas tree lights because if one goes out the whole string won't light.

Example:

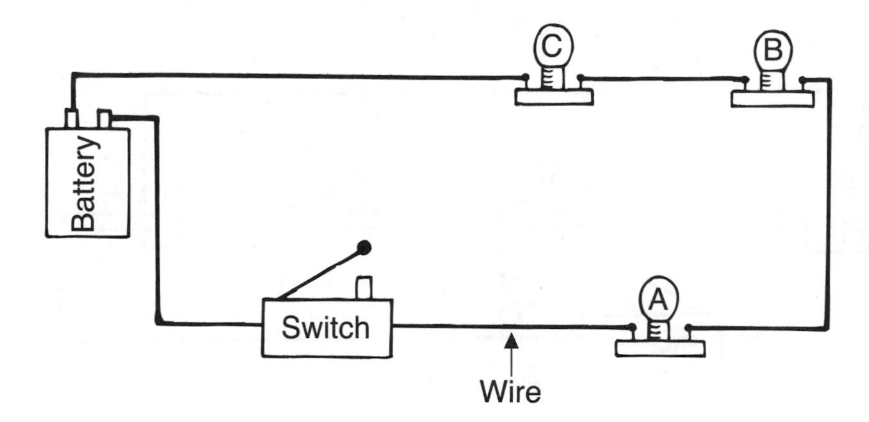

Parallel Circuit (page 108)

1. Yes, because there is still a completed circuit on another path.
2. The circuit won't work. The circuit will work because it is complete.
3. The lights will be brighter.
4. Yes, because if one light goes out, the rest still stay on.
5. Answers will vary. Possible answer:
   In our car there is a parallel circuit because if one headlight goes out, the other still works.

Example:

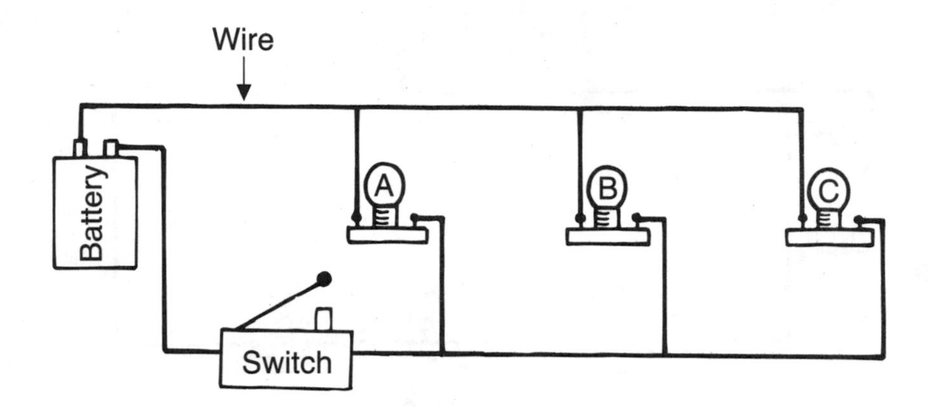